E-BOOK REVOLUTION

E-BOOK REVOLUTION

THE ULTIMATE GUIDE TO E-BOOK SUCCESS

EMILY CRAVEN

Craven Publishing

Australian Edition

© copyright Emily Craven, 2012 (Edition 2 – 2019)

This book is licensed for your personal enjoyment only. The information and advice in this book are suggestions only and before you enter into any contracts or financial payments you should make your own investigations and judgments based on your situation. This book is sold subject to the condition that it shall not, by way of trade or otherwise, be resold, hired out, or otherwise circulated without the author's prior consent. Thank you for respecting the hard work of this author.

Please note all spelling is in English (Australia).

Contents

Free E-book Copy   ix
The Revolution: An Introduction   xi

Umbrella 1: The Pros & Cons of Publishing Vs Self-Publishing

1.1 The Reality Of Publishing	3
1.2 Niche Markets – The REAL Reason For Manuscript Rejections	5
1.3 The Global Numbers Game - Traditional Vs. Self-Publishing	9
1.4 E-book Opportunities For Newbies, Old-bies & Anyone Who Wants To Take A Crack At It	11
1.5 E-book Opportunities For Established Authors, Agents, or Publishers	17
1.6 Will Self-Publishing Ruin Your Chances?	21
Umbrella 1: Important Chapter Links	25

Umbrella 2: Creation & Publication

2.1 First Steps To Success: Preparation	29
2.2 There, Their or They're? Self-Editing	33
2.3 Are You Living In A Little Pink Cocktail Umbrella Heaven? Doing Market Research	39
2.4 The Foundation For Published Success: Pitches And Proposals	45
2.5 It Started With 'The'	51

2.6 Got It Covered?	53
2.7 Formatting, Decrease The Number Of Grey Hairs	55
2.8 Avoid the Meat-grinder! Creating an Epub File can be as Easy as Writing a Blog	61
2.9 How Non-US Authors Can Avoid Paying 30% Tax To Smashwords and Amazon	65
2.10 The Superbook	69
2.11 If You Want To Be A Great Writer, Educate Yourself	71
Umbrella 2: Important Chapter Links	75

Umbrella 3: The Hype – Marketing

3.1 Marketing Starts Yesterday	79
3.2 Websites: How To Entice Readers Into A Buying Extravaganza	83
3.3 Turning Your Once Off Readers Into Return Readers	87
3.4 Reviews: How Credible Is Your E-Book?	91
3.5 Handling Feedback Is Like Juggling Tables	97
3.6 You May Have The World's Greatest Book But, If No One Reads It…	99
3.7 The Road To Free Traffic Is Paved With Articles	105
3.8 Consider How The Humble Short Story Can Increase Your Fan Base	109
3.9 The Power Of Recommendation: Contributing To Your Community	111
3.10 You Want Raving Fans, Not Rabid Ones!	115
3.11 Facebook and Twitter: The Gateway To Thousands Of Readers	119
3.12 Go Viral With YouTube	123
3.13 Advanced Techniques For Generating Demand	127
3.14 Would You Promote My Book? Pretty Please?	129
3.15 Want To Charge Top Dollar For Your E-Book?	133
3.16 Ten Shades of Author Collaboration	139
3.17 Webinars: Connect With Readers Across The Globe & Generate Massive Sales	143

3.18 A Writer's Money Isn't Just In The Books	149
3.19 The Business of Being a Writer	151
3.20 Why Give Your E-Book Away For Free?	155
3.21 The Ultimate E-Book Launch	159
3.22 How To Get Your Books Into Your Local Book Store	161
3.23 How To Structure An Author Talk	175
3.24 Crowd Funding 101 For Writers	181
3.25 Interactive Storytelling: Real-Life Choose Your Adventures	187
Umbrella 3: Important Chapter Links	195
The Revolution: An Ending	199
Have You Found This Book Invaluable?	201
More Books By Emily	205
About The Author	209
Contact Emily Online	213

Free E-book Copy

With This Book You Also Get A Free E-Book Copy!

Just email emily@cravenstories.com with the following to prove your purchase:

- A photo of yourself holding the book PLUS
- A photo of your name written in pen on the copyright page
- Mention where you bought the book from.
- What format you would like the book in: .epub (iPad, Kobo, Nook etc), .mobi (Kindle/Amazon) or PDF.

Once Emily receives your email she will personally email you back your e-book copy. She's only human, so give her a couple of days to get to your email. Thank you once again for purchasing the book!

The Revolution: An Introduction

There is no use kicking and screaming like a 2 year old in the throes of a marathon tantrum – the e-book revolution is here to stay, for better or worse. While every author's ultimate dream is to sniff the glue binding of the crisp, off-white pages of their first glossy masterpiece, the reality is that most of us will never write a book that exactly fits a publisher's idea of a mainstream, marketable novel. However, with almost nine million searches per month for 'e-books' on Google and 2.5 million searches for 'free e-books', we would be out of our minds not to capitalise on this digitisation of our industry.

Now, I'm one of the first to admit I'm a bit iffy about reading a novel on a screen. The weight, the buttons, the multiple formats… it just doesn't seem, well, *natural*. But with the fall of bookstore giants like Borders and (in Australia) Angus and Robertson, and the gates of publishing houses firmly closing on unpublished authors, it's ridiculous to believe the old model of getting published is going to work in this digital age. So I decided it was time to part the fingers I had been holding vice-like across my eyes and start figuring this mess out.

And it IS a mess. All new and scattered, the information is there IF you have a lot of time to trawl through it and you look in the right place. Fortunately, I have been lucky enough to have time and direction. I have researched and gathered, joined courses which have cost the better part of three around the world plane tickets, and come up with several bags full of golden notes. But as all writers know, a bag full of inspired notes is about as useful as a Mardi Gras parade without neon floats bigger than your house; without organisation (or parade floats…) it just doesn't make sense.

So this book is all about making sense of the madness and taking you through one piece of the puzzle at a time. I want you to learn as I learnt and to discover as I discovered, because let's face it, we all need a path and without a light to guide us we are lost. This book will cover everything, from readying your manuscript, to e-publishing, to marketing and social networking. Get ready writers, the e-book revolution is here.

Umbrella 1: The Pros & Cons of Publishing Vs Self-Publishing

1.1 The Reality Of Publishing

Download the audio for this section at: *http://tiny.cc/audio1pt1*

It's time to tear away the grand illusion of publishing like a Band-Aid stuck to a hairy leg. Folks, this is the reality of the industry as it stands. Hopefully this will free your mind to the possibility of making good money from selling your e-book online.

The publishing industry is going through a forced (and much needed) overhaul. The way some publishers talk, you'd think they were being held at the literary equivalent of gun point! One thing is for certain: many challenges face the traditional publishers to whom we turn to release our long-nurtured work. Traditional publications are expensive to produce, promote and distribute , which results in an average book cost of $25. This cost is passed onto the consumer, and let's face it, people need to be pretty convinced to spend $10 these days let alone $25. While publishers have the luxury of dictating what they believe the public will read this more often than not results in difficulty in predicting demand.

Publishers cannot afford large budgets to promote the majority of their list which leaves many authors to be their own publicists. The feeling that you had made it and could relax? Yep, that goes up in smoke, because if you don't help speed the sales along yourself then your book, your labour of love, is going to have a much shorter life than you thought.

Novels only have a limited geographical distribution, generally restricted to the country in which they are published. Shelf space in bookstores is limited and a book has only three months on the shelf to prove its worth against the other midlist books before being returned to the publisher and pulped (destroyed). There are no more print runs of your book and that is the end of it. This is a danger for every published author. For example, one of Garth Nix's famous books, *Sabriel*, only sold **just** enough copies to keep getting small print runs. It took seven years until it had gathered enough popularity to be placed prominently on bookstore shelves rather than shoved in a back corner.

Right, there's half the Band-Aid gone, now to how authors make their money. If this has demoralised you, just take a deep breath and rip off the rest with me. The hair will grow back...

One of the burning questions we all have is, are e-books profitable? It appears through all the research I have done and successful e-book authors I have talked to that e- books can be *more* profitable for an author than traditional runs.

Let's take a look at the traditional publishing mouse wheel. A first time author is lucky if they get a $1000 advance. As this is an advance against royalties you must make over $1000 worth of royalties to start receiving any more royalty money. The average royalty payment for an author is about 40 cents per book. This means you must sell 2500 copies to make your $1000 and start getting paid ANYTHING. To get another $1000 in royalties you must sell 5000 copies in total.

If this book is sold for $15 a book (we're going lowbrow here), they make $75,000 for 5000 copies and you get $2000 of that. That's 2.6% of the total earnings. For potentially several years of work by the author, that's a little steep. Most first time books will sell fewer than 5000 copies. The only marketing done for first time authors is generally a mention on the publisher's website and/or newsletter and a local book launch. That's it. How your novel sells is entirely dependent on where the bookstore owner decides to put it on the shelf. It's not based on keywords or searches but biased opinions about which books will sell better than others.

Now let's look at the profit when selling an e-book. Even if you sell your book online for $7 a copy and you only sell 2500 copies you still earn $17,500 – a damn sight more than you would through a traditional publisher.

Are those ripped follicles starting to feel better now?

1.2 Niche Markets – The REAL Reason For Manuscript Rejections

Download the audio for this section at: *http://tiny.cc/audio1pt2*

At some point in our writing careers, all of us have felt that our novel is a masterpiece. Some, more than others, will be of the opinion that they are the fricken Michelangelo of the written word. Then the rejections roll in... and in... and in. Many of these are due to the manuscript (MS) not being of a certain undefined standard at the time of submission (or, God forbid, we aren't as good at crafting our Sistine Chapel as we thought). However, a significant number of MSs, particularly non-fiction, are rejected because a publisher just doesn't see a large enough market or simply that particular subject does not resonate with their tastes. This is a fair enough decision, as money has to be made and redundancies kept in check. Not every MS can be accepted, and even if a novel does make the cut, the publishing houses cannot afford to extensively market all the books they sell. Unfortunately, over the past several years the number of publishing houses has decreased and fewer authors are being published.

However, publishers **do not know** all markets and fan bases available or how many people participate in these specialised niche markets. As such they are not equipped to serve all authors in this digital age. A publisher with 30 years of experience is not necessarily going to get the popular new trend of Crimping or the hundred thousand fans around the world who play *Magic – The Card Game*.

Currently, readers are denied the freedom to discover new voices, yet we are entering a future where more people will read what they are interested in and discover fields they had never previously considered. The opportunities provided by e-books to reach these niche markets are immense. Briefly, a niche market is a market that is fixated or interested in a particular topic. For example, I have written a non-fiction novel on Gap Year travel, which is a 'niche' of the travel genre and a phenomenon that is hugely popular in the UK and US but next to unknown in Australian culture.

A clear example of publishing houses' inability to serve all niche markets is shown by Mark Coker, the creator of Smashwords, a free e-book publishing site. Mark began this site after his struggle to traditionally publish a specialised soap opera novel called *Boob Tube* (really, who could resist??). Represented by the well-respected New York City agency Dystel & Goderich to top publishers of commercial women's fiction, Coker's MS (co-written with his wife) was thoroughly rejected. The pair undertook a major revision of their book using the comments of the publishers, then they shared the manuscript with test readers interested in the soap genre and received an enthusiastic thumbs up. When they resubmitted their MS, still being represented by a top firm, they were rejected. The market was too small to interest mainstream publishers. The struggle sparked an idea for Mark and Smashwords was born.

There are many brilliant writers out there who have an interest in thousands of topics. I'm sure many people have no idea what a flash mob is, but a book on how to organise a flash mob with ideas for start-up groups to try could be a potential best seller to interested readers. After a quick check, I found there are over 550,000 searches for flash mob or flash mobs per month on Google. (For some fantastically entertaining videos look up flash mob on YouTube.)

Internet marketers have been capitalising on niche markets for years. Brett McFall commissioned a ghostwriter to create a book on making money with scrapbooking. McFall made $40,000 in a year from the book. The ghostwriter wrote it in 14 days. Imagine what the niche market potential is for someone such as yourself who actually knows how to write and has been working on their novel for years. It is all about pitching your novel to these markets correctly, which we will cover later in the book.

E-books provide an opportunity to reach niche readers where an author knows the needs of their smaller group of enthusiasts better than a publisher. We all place such high hopes on publishers but they are just people like us, wandering around their own little world preferring red over black, sex over sleep and who just do not want to read another teen romance no matter how much it is like *Twilight*, in fact, **especially** because it's so much like *Twilight*.

1.3 The Global Numbers Game - Traditional Vs. Self-Publishing

The internet is so wide reaching (heck, even the Pope has a Twitter account) and it's important to note that the benefits and successes of e-books are not restricted to niche market books. Traditional commercial fiction can sell just as well with this method, if not better. The only difference is the stiffer competition within the commercial fiction genres. An author will have to work harder to market their book, but if they are savvy and can connect with their readers on another level, they can break through. This book is largely about how to do this.

Now I can hear the worriers chattering away in the background like teaspoons in a blender, asking "Won't e-books with narrow markets face the same problems as hard copies?" It's true; there are only so many people that may want to read about a specialised topic.

So, let's return to the global numbers game and say 20,000 people enjoy this niche market. What are the chances that all those 20,000 people around the world are going to find your book, in an Australian (or a UK/US) book store within three months of its first release (remember our works have a three month hourglass here)? Guaranteed not even 1%. These books will *never ever* get traditionally published, not if the best you can get is 200 books sold in those crucial three months. If they are published in Australia, they will never get the chance to be marketed to the people in America, the UK and elsewhere.

But the book has still been written, so why should it never be published if it has information that someone is going to find helpful and buy? And that person is not in Australia? Are you going to just jot down the last several years (or decades) as time wasted and move on? Or cling to your manuscript like a cat lady clings to her spinsterhood?

The best thing about the web is it doesn't matter what country the person is in, you can reach them. It doesn't matter if someone discovers an interest in your topic two years after you have released the book; your book is still on the web so you can still get money from it years later. Traditional publishing is restricted numbers wise, time wise AND geographically. You may not be an international star but at least you will get money for your hard work.

One great article on the reality of publishing is by world-famous author Ian Irvine. "The Truth About Publishing" is a must-read from a popular author who knows exactly what goes on: *www.ian-irvine.com/publishing.html*

1.4 E-book Opportunities For Newbies, Old-bies & Anyone Who Wants To Take A Crack At It

Download audio for this section at: *http://tiny.cc/audio2pt1*

Unlike traditional publishing an e-book costs nothing to produce and nothing to place the book online to sell. It almost sounds like a badly written sales pitch, 'No capital? No worries! Living below the poverty line? It won't cost you a cent'. In a horror movie it would end with something along the lines of, 'Just your soul….' But there really are no hidden skeletons with e-books; this becomes quite clear once you've done your research.

Reputable distribution sites exist, such as Smashwords, or Draft 2 Digital, which allow an author to produce their e-book for free and market it in major online bookstores such as Amazon, the Apple store, and Kobo. All rights are retained by the author and if an author decides to remove their book from the Smashwords or D2D site they can at no cost. All these sites ask for is a 15% commission for selling through its site and those of the online stores it distributes to. This means the author receives a higher commission of 85% per novel as opposed to 40 cents per novel received through a traditional publisher. You can crush a traditional publisher's e-book with prices a fifth of what they charge and come out with more money. The reality is that if you don't do it in your niche (or your genre) others will. There is no use standing in the bull ring with a fluttering flag when all your competitors have scaled the wall above you.

Okay all you tight lipped balls of negativity – spit it out. You are dying to say that a site that charges 15% commission is NOT free. At the final sale this is technically correct. However, it is free to PRODUCE your work; you can set it up in the right format, have a cover put to it and then save it in several different formats. This is all free! It does not cost you a cent, a penny, a wooden bead in cash or barter items to produce this. (You would be spending large amounts of time on the book anyway if you were doing it through a traditional publisher.) You have already done the hard work revising and editing; it only takes a couple of hours to make sure that the words look right on the page and you don't print the font so small that readers require magnifying glasses.

If you do not wish to keep it up on these distributor sites you can remove it from the site and go. Then you can reformat and sell it yourself or give it to your friend's, aunt's dog groomer, it doesn't matter. However, you cannot take the formats Smashwords and others convert for you from the site and sell it elsewhere (as stated in their terms and conditions). For one, they are not a conversion service, they are a distributor; and two, nicking off with the conversions and selling them elsewhere is just plain rude. I'm sure they won't begrudge you giving it away to your aunt's dog groomer though… The point is, unlike a traditional publisher, if you are not happy with it you can pull the e-book from the site and try something else. What that something else is I am uncertain, but if you feel like cutting off all your distribution arms, it's entirely up to you.

It does not cost you to print the books, it does not cost you to ship them to the book stores and it does not cost you to hire an illustrator. (Though you should have an illustrator if you are serious. Just ask Matthew Reilly.) You pay no upfront costs, unlike a traditional publisher who is putting their money on the line to publish your book. Though the author doesn't have to pay, the publisher has spent all their money on printing and suddenly your marketing budget is looking a lot leaner… With no upfront costs you can self-publish and not be out of pocket.

Initially self-publishing meant you went through Lulu.com and they printed out 2000 copies of your book for $5000. Those books then sat in the author's garage, gathering a veil of dust and mouse droppings. Without the necessary connections, the author could not shift them. Then the author was left with 2000 nice paperweights AND no money.

2.1 E-book Opportunities For All

This scenario is not true of e-books because they are digital files that can be distributed through the web. A commission only comes out when you get paid for a download. In traditional publishing a publisher takes a 90% commission, whereas these e-book store websites only take a 15-30% commission. How much more does the author get? That's a lot more per sale.

These online bookstores are needed to sell a reasonable number of copies of your e- book, just like you need a traditional bookstore to sell hardcopy books. You can try to do it yourself off your own websites, but this takes a whole new set of knowledge to set up (some of which we will learn in our marketing section). People go to the online bookstore because they know that is where they can get books! By selling your book only from your website you are amputating your leg and throwing away the crutches. With an e-book store it is a win-win situation; you do not have to pay thousands of dollars to print and distribute the books or to hire people to do the layout and formatting. You are placed on a site (like Amazon) with high traffic that only takes a 30% commission from books that are sold (not ones sitting in your garage at home) rather than the 90% commission taken by a publisher.

Earning 70% more from my novel? Now THAT'S a good day...

***

Whenever I talk about e-book publishing the main complaints are not about commissions or free production; they are about TIME. Time, time, time, time, time. Several people have scoffed into their coffee mugs, "Costs me nothing? Nothing my arse! All that time I put in, all the food I've eaten, all the toilet paper I have used! That's not nothing! That's a whole lot of extra work for <insert profanity here>."

As a fantasy author (mainly) I hate visiting the halls of reality, but in this case it needs to be done. How much time have *you* spent approaching publishers, agents and sending out submissions? Is this not a whole lot of extra work for nothing? Please show me this magic carpet of denial. Producing an e-book does take work but you would be revising, editing, rewriting and marketing as a first time author of a traditional publishing contract ANYWAY. But a contracted author would only get paid $1000 for it. That is not a lot of money. In terms of little profit, you get $2000 for selling 5000 copies of your print book. So although it is argued that making an e-book costs time (if not physical money) it would cost time no matter what route you took for little monetary return.

Truth be told, the author has already spent hundreds of hours of work developing their novel before they have checked to see if it has a market. Initially, you write it for you, and then you write it to help others, then you think about making money off it. That's the creative process (though writers would find that if they did this process in reverse it would be more profitable in the long run. Thinking what the market wants/needs FIRST and then giving it to them will see you more successful than most commercial authors). The author, once they are finished, is invested in the project and does not want to see it take an extended holiday on their computer's hard drive. The author has already made a mammoth effort in writing the book, so what is a week or two to do the production properly? If you aren't willing to do this work, then you do not want it badly enough. If you don't put the effort in, why should anyone else put the effort in to read it?

You do not have to have your book professionally edited. A manuscript can come a long way by having other writers workshop it or by being read by someone with a good grasp on grammar. However, **do not** be the only one to read your novel before you publish it! You are too close to the piece. Get honest people to give you honest feedback. Family and friends rarely count; they have to live with you after they give the feedback. They don't want you staring daggers into their chests because they told you your main character was a sleaze. If you do want your book polished, local writing centres offer affordable editing.

Online communities like *www.fanstory.com* or *www.wattpad.com* are also essential to an author if they wish to grow but don't have the money for professional courses. If an author refuses to take the feedback offered by other writers and their sales suffer then that is their business. There are always people who are too precious about their work and in the long run, publishers will not work with those people anyway.

Is everyone okay now? No more tiny voices conducting a UFC match in our brains? Wonderful! Let's move on.

2.1 E-book Opportunities For All

E-books are **convenient,** folks, and everyone loves convenience, particularly America. Exercise? Overrated. Browsing without a search engine? A time-waster. The appeal of purchasing and reading a book immediately rather than journeying to a busy shopping centre or having to wait weeks for an order is becoming more apparent to online consumers every day, as shown by the number of searches for e-books on Google (almost 2.5 million searches each month).

Are you not getting published because you don't fit the normal formats? An e-book is not restricted to a novel; it can be used to sell short stories, novellas, poetry, newspapers and magazines (literary or otherwise). I have no doubt a new trend in the future will be 'entertainment bites', sections of prose, jokes or whatever takes your fancy to help reduce the friction burns caused by commuters dragging their behinds to work every day.

But by far the most beneficial point of publishing your work as an e-book is that these books **never** go out of print, **and** they can be sold to people in every country in the world with an internet connection. The singular three month window of opportunity simply doesn't exist online. Online, money can still be earned long after that established bookstore deadline. A book can sell 24 hours a day for a year or ten, to hundreds or thousands of readers a year, without being pulped. With such 'low numbers' of readers (worth thousands if you market it right) a publisher would not even consider accepting such a book for their list.

Aren't you glad you found this book?

1.5 E-book Opportunities For Established Authors, Agents, or Publishers

The e-book revolution not only benefits first time authors but also publishers, agents, authors who are already in print and even a marketer who has hired a ghostwriter.

Popular Authors

Can produce their own works and use their loyal fan base and websites to market their e-books along with such tools as social media and webinars. In this way they can make far more money than they would through publisher royalties because they know their readers are clamouring for their books. Add your own print on demand copies and you can cover both traditional and digital worlds.

Cory Doctorow, a science fiction writer and New York Times bestseller, is one such author who has capitalised on this phenomenon as I'll expand upon later. Popular authors have ways of pushing the boundaries of the written word and getting a large amount of feedback on what does and doesn't work. An author can launch a book worldwide on the release date through a webinar. They can charge for talks, specialised writing courses and virtual tea parties without having to leave the comfort of their home to do it. All this I will expand upon under Umbrella 4.

Established Authors

Who have penned a novel in an unrelated genre to that in which they have been published can also avoid jumping publisher hurdles by publishing their work as an e-book. You're a thriller author but you've penned a children's story about bunnies? No need for that work to go unseen. Do you have a novella or a short story but you are not 'popular' enough to have them considered for the big name anthologies? You can release them yourself and not only make money but also build your reputation.

Agents

Can use the facilities of e-book publishing sites to promote the work of their clients who retain the digital rights to their out-of-print books. An agent who knows a little internet marketing can generate a lot of extra income for their clients who may have fared badly in their initial three month launch period but whose work still merits attention from the rest of the world (This in turn generates more income for the agents!).

Marketers

Even marketers who have a ghostwriter to pen their idea can use a distributor site like Smashwords or Draft to Digital to convert their PDF or Word document into every single format imaginable so they can cover all e-book bases and distribute to online e-book stores. Private white label books are not permissible for upload into most e-book platforms and the book cannot be constructed of public domain material (eg. a compilation of 40 articles freely available on the Ezine website).

However, if you are an expert in your field, there are great ways to use e-book platforms to your advantage. Though affiliate links are banned as part of most distributos terms and conditions, if you are an expert in your field and can genuinely pen a good-value and reasonably sized book, you are permitted to provide hyperlinks at the end of your e-book to your other books, website or Facebook fan pages and Twitter accounts etc.

Publishers

Though they appear yet to do so, can also capitalise greatly on the e-book revolution. A publisher working through online bookstores can receive up to 85% of the net sales price of their authors' e-books and access a wider distribution than just their personal website. Publishers currently receive only about 40-50% of the price of a traditionally

2.2 E-book Opportunities For Established Authors

printed book (due to distribution and author costs) or even only 20% if books are returned from stores. But digitally these 'middle man' costs barely exist. Many are reacting just as the music industry did all those years ago and not looking for ways to embrace and use the technology of the day to generate as many, if not more, sales than before. It is far more beneficial to work IN the internet machine rather than against it.

Publishers also need to consider that the internet is a fantastic place for them to test the demand for a new author's novel before committing to a physical print run. Do you have a novel that is fantastically written but wouldn't normally be published due to the economic concerns of printing for a smaller fan base? Or perhaps the novel is not a genre or style that has been traditionally published before and you would like to see if it is received well by the public?

All of these scenarios can place a publisher at the frontier of edgy book culture without impacting too much in terms of printing costs. Professional online publishing is enticing to authors because they get to experience the professional publishing process while possibly receiving a higher commission. Also, more of the book's budget can be spent on marketing to allow the author greater exposure for their first novel so it can prove its worth. The internet is the cheapest place for a book to fail.

In fact, several of the marketing strategies in this e-book can be applied by publishers to increase interest, launch books and create strong communities around their authors with barely an advertisement in sight.

Did I not promise this book had something for every angle?

1.6 Will Self-Publishing Ruin Your Chances?

Unfortunately with any new and lucrative opportunity comes great concerns. Very few people can say that they haven't shared or 'pirated' movies or music in their lifetime. If we were to express our piracy in costume, almost 99% of us would be wearing eye patches and calling everyone 'Matey'. Arrr, it's the truth. Over several years we have been bombarded with stories of the music and film industries' apparently declining sales and battles over copyright.

The chief concern among authors is losing money due to the sharing of their e-book files as opposed to the purchase of them. One solution to this is the implementation of digital rights management (DRM) codes which when placed on e-books 'prevent' sharing. Yet, in the long run, these sorts of prevention methods are counterproductive because they treat lawful customers like criminals. What we all need to ask ourselves is this: is it worth it penalising a fan for promoting our work? Many consumers resent DRM as it limits their ability to fully own and enjoy their e-book. Readers are in fact actively rebelling against DRM.

There is a growing body of evidence that authors and publishers who have abandoned DRM are actually enjoying greater sales of their works. As noted by Cory Doctorow, it is practically impossible to prevent copying in this technological age. He therefore markets his books by giving away the e-book version (under a commons license that allows non-commercial sharing) to attract readers to buy hard copies of his works. Essentially, by enlisting his readers to promote his novels, he increases his sales by simultaneously releasing the e-book and published copies. It is worth noting that Cory has had his novels on the New York Times bestseller list over the past couple of years.

Now, I'm not suggesting you just suck it up because you're not going to earn any money. Nor am I suggesting you close your eyes and give your baby away for adoption to the nearest pirate-clad stranger. Currently in bookstores we are seeing a decrease in the number of books sold due to price, however, a digital e-book is cheaper and with the correct marketing and pricing people are inclined to buy this non-expensive entertainment as opposed to pirating it.

As a way to visualise this copyright issue, think of digital sharing as a person lending their friend a physical book. That borrower does not want to buy the novel of an author they don't know, but by their friend lending the book to them they are also being given a recommendation of the quality of that work. As a result the borrower may become hooked on that author's work and now the author has a new fan and customer who will recommend the book to others.

Word-of-mouth and sharing are some of the most powerful ways to create a fan base and hence income. Cory has almost 111,000 twitter followers; it's ridiculous to think with 111,000 fans you would make no money. How did he get that many followers? Because he got his work out there by giving it away and now he is reaping the rewards.

Should an author be worried about the sharing of their work, or should they be more concerned about their work of art never, ever, seeing the light of day as it is rejected year after year? Personally I would prefer not to waste years in which no-one reads a word of what I have written, no-one follows my writing and no-one knows who I am. Initially, it is more important to reach an audience than it is to make money. Hell, it's like a comedian refining their jokes until their big moment at the comedy gala and never getting there because they didn't do any damn practice shows! No one knows they're funny, and unless they shove their material in front of people they will never get the attention they need to make it to that gala.

Neil Gaiman, author of *Stardust* and *Coraline* and one of the world's most beloved writers, has also spoken about how piracy may even be beneficial to an author, as seen on his video on YouTube (*http://tiny.cc/NGcopyright*).

In some cases, showing you can reach an audience is beneficial. This relates to the concern that first publishing rights for a novel will become less valuable to a publisher if it is already digitally published. This is possible but not inevitable. You have to consider a publisher's view on the commercial marketability of your work. If you, as an author, can prove that you can sell books and that a market exists for

1.6 Will Self-Publishing Ruin Your Chances?

your writing, then you increase the value of your body of work. You will also have a fan base the publisher can market to. Scott Sigler, another New York Times bestselling author, created a large online following by giving away his self-recorded audio books as serialised podcasts, similar to an old-school radio drama. This was all *before* he was offered a traditional publishing contract. By using the internet to his advantage, and leveraging his books and the art of viral marketing to his fans, he was able to prove the quality of his work to publishers. This is an alternative for people who can't bear to have their book in a possible-to-pirated digital form, but want their work to be known and appreciated.

Another common question is how the quality of e-books is maintained. Aren't e-books a medium where anyone can publish the drivel they have written? I think we can all agree we have read books by highly regarded publishing houses that have turned out to be utter crap. I mean, you are damn near ready to burn that book and recommend it to your worst enemy because of the time it wasted. I continuously ask myself how some books made it through. In the digital world, like the physical world, one thing can be certain; it is up to the readers to decide what is worth reading. The great authors bubble to the top through word-of-mouth, and the lesser authors drop out of sight. There are ways to make sure that your MS is up to a high enough standard so that your writing does not fall by the wayside due to easily fixed issues.

Umbrella 1: Important Chapter Links

Audio

- 1.1 The Reality of Publishing: *http://tiny.cc/audio1pt1*
- 1.2 Niche Markets – The Real Reason For Publisher Rejection: *http://tiny.cc/audio1pt2*
- 1.4 E-book Opportunities For Newbies, Old-bies & Anyone Who Wants To Take A Crack At It: *http://tiny.cc/audio2pt1*

Other Resources

- Smashwords: *http://tiny.cc/smashwordsEC* (affiliate) or www.smashwords.com
- Ian Irvine – The Truth About Publishing: *www.ian-irvine.com/publishing.html*
- Online Feedback/editing community: *www.fanstory.com* and *www.wattpad.com*
- Digital Guru Cory Doctorow's bio: *http://craphound.com/bio.php*
- Neil Gaiman copyright video: *http://tiny.cc/NGcopyright*

Umbrella 2: Creation & Publication

2.1 First Steps To Success: Preparation

So what are the steps to success when, having weighed the pros and cons, you decide to digitally self-publish your book? Have a drink? Granted, big leaps of faith require a drink or two, but no. Brag on Facebook? Don't count your chickens before they hatch. Start scheduling a string of very important black tie speaking gigs? Hardly, you are not the fricken Michelangelo of the written word just yet.

Most importantly – please never forget this – is to write a shit hot (great) book. Then revise, edit, revise, repeat, get five different friends to edit, hire a professional, revise, repeat, read backwards, forwards, out loud, revise, repeat. One of the major pitfalls for self-published authors is not having the manuscript thoroughly dragged through the bull pit in terms of accurate grammar, a thrilling plot and character development.

Any writer who is truly serious about producing good content should hire an editor. (If you want your readers to give a damn about you, you have to give a damn about them!) If you cannot afford for them to do your whole manuscript I would recommend that you get enough money together so that they can look at the first 30 pages of your work. Any editor worth a grain of sand will be able to tell you most of your repetitive writing boo boos from the first 30 pages. They will be able to tell you if you have fondness for a certain word, long sentences, cliché characters, superhero dialogue or always writing there/their/they're as 'there' regardless of the correct form. The editor needs to explain to you exactly *why* they are making these changes; if you don't learn from the experience then you have wasted your money. From an editor looking at the first three chapters of my work I have improved my entire manuscript by at least 200% – that is how powerful this is. Writers' centres also offer appraisal or editorial services for a very low cost to members.

Not only should you be making this your number one priority but you should also take at least one course on editing or barter with fellow writers for a read through. Your friends and family do not count. They are completely biased and entirely too nice and, in most cases, do not have a clue about what they should be looking for anyway. Instead having a solid writers' group with different tastes who read different genres will see your manuscript jump leaps and bounds without an editor, because each writer has a different editing focus.

Now don't start giving me these predictable excuses like, "What if someone steals my idea?" If you think this way then your manuscript will in all likelihood only reach a hundredth of its potential and, more often than not, will never see the light of day. You can have the world's most fantastic idea but if you don't share it, it's not going to mean a damn thing. If you do not gauge the reaction to your piece before it goes out into the big wide world of critical readers you will be in for a rude awakening. More of a ripping-your-heart-out awakening as opposed to a pleasant splash of a bucket of water over your head. I know many of you don't want a single cell of bacteria, let alone a human being, look at your work without it being perfect. However, if you wait for perfection you will wait forever. Writers' groups are aware and expect that every piece of writing you give them is a work in progress. The sooner you beat your bad writing habits out of you the better.

Online writer and editor communities are also a fantastic place to learn the art of editing your work. While you have to weigh every comment for its relevance, you also need to consider that there is always a reason behind every comment or suggestion people make. If more than one person makes the same comment you know it's an area that needs looking at. At all times, keeping an open mind is key. One such online community for chapter segments is *www.fanstory.com*. In this community it always pays to actively read through other people's work as well for two reasons, a) for politeness and b) so you can see the mistakes other people make, see the styles other people use and view well-written work that can help you improve how you write. Also, by taking an interest in other people's work they will naturally take an interest in yours. The community is a barter system and to receive you must also give. Another community is the Wattpad community. Wattpad allows writers to upload manuscripts, participate in writing groups and have readers give feedback, make comments or annotate their work. The only thing it appears to be lacking is the ability to draw cartoons in the margins.

All these elements are invaluable when making sure you have a quality product to release to your market. Wattpad also acts as a conduit for publishers to search comments and discussions for manuscripts they wish to publish – because yes, this internet writing community has just launched their own 'traditional' publisher. If that is the avenue you are more inclined to pursue, rather than self-publishing, then this site is doubly for you. As with any community you need to make sure your manuscript fits in with the sort of genres that participants are interested in. Wattpad is a publishing community of fiction and fanfiction. As stated by Wattpad, "We empower diverse voices and stories all over the world, helping you build a community of readers." So attempting to post your paranormal romance involving vampire rabbits may work here, but not on other sites.

Dozens of online communities abound (such as Scribophile, Critique Circle, or even Reddit). Here, other members provide critiques for brownie points in exchange for people editing their work. You can get a lot of value out of free memberships to these websites (though of course there is always a paid membership version). What you can learn from such minute editing is intense and well worth the effort in your final product. In saying that, your book, and your independent author career will always gain a much higher trajectory if you find an editor who can support you through multiple series and books, lending a consistent high quality none of us can achieve on our own (even with Beta Readers!). But these online communities will, in the beginning, help you to hone yourself editing skills so that when you DO hire an editor (like myself!) they will be able to provide you a cleaner edit, faster (and cheaper) than if you dumped your first draft on them.

Self-editing is the most crucial skill a writer can learn and it applies to any writing form: short stories, articles, poetry, haikus, everything! Every piece should be taken through the ringer by yourself and other people before you even consider publishing it. We will go through the basics of self-editing in the next section.

2.2 There, Their or They're? Self-Editing

Download the audio for this section at: *http://tiny.cc/audio3pt2*

So you are done. You're over the moon, you're doing back flips off the wall, cartwheels through the living room and you're bouncing about like you've had ten cups of coffee and one giant wake up pill. The manuscript is done, the characters have had their last dance, ninja kicked their last bad guy, breathed their last breaths and you can relax. Right? I think you know the answer to that one.

The number of people who have told me that revision is key are too numerous to count, but one thing's for certain: they are right… the bastards. Because good writing is re-writing. However, there is a slow, chaotic way of editing, kind of resembling a 3D Picasso jigsaw, and a methodical way of doing it which also keeps your sanity firmly rooted to planet Earth. The first step is to see revision as the fun part. It's easier than creating something from scratch, no accusing cursor demanding why you haven't burnt the suspicious pair of underwear in the cookie jar yet. Now that you know the whole story yourself you can go back and make sure all the hints you thought you included are actually there and not just figments of your imagination.

Self-editing is daunting; where do you start? What are you looking for? What if you can't figure out what you're doing wrong? Don't panic! As I said, there is a simple way and that involves cutting editing into small manageable chunks by searching for a different thing in each read through. To help calm yourself you should set aside your writing for a period of time. The longer you leave it the better. For those of you who are so addicted you can barely bring yourself to switch off the computer, try to hold off for a week at the least. Then let the fun begin.

Reading 1

Do a non-stop read-through of the entire novel on a day when there are no distractions. No kids with colds insisting you try the muffins they've made, no topless partners walking through your work space, nothing. Read aloud if you can because that helps you pick up any awkward points more easily. Have a pen or highlighter in your hand and every time there is a stumble, discomfort or something just bothers you, make a mark on the page.

But keep reading!!!! Then come back later and decide what troubled you. Did you put a comma in the wrong place? Was the sentence awkward? Did you slip from the point of view of the granny to the wolf? Or have a nun say something that would come out of the mouth of a stripper? As a first stab in the dark, do what you can to sort out the marks on these pages.

Reading 2

Here you need to check structure, particularly whether the order of the story, tension and pacing makes sense. Should you really be building the tension of your character buying the sandwich and only spend one line on them getting robbed in the alleyway by a midget? Here are several things you should look for:

For non-fiction: is everything in its logical order? Are you giving enough information to lead onto the next chapter? Does the end of one chapter make you want to read the next? Do you do a brief paragraph at the start of each chapter listing the reader benefits of that section to draw the reader in? Does the introduction grab attention? Is it emotive, does it identify with the reader's problem? Does it offer the promise of a solution? Does the introduction tell the reader how this book is different, why it is the one that holds the answers?

At the end: what is the conclusion you want them to draw? Gently lead them to it and piece together all the previous chapters painting the big picture in one short, sharp ending.

For fiction: how is your beginning? You only have one first line and one first paragraph. It is at this point where the reader is going to decide whether you are worth investing time in. Does it set the demeanour for the rest of the novel? If you are not happy with your beginning, try to start from the first interesting moment in your story; start in the action at a point of conflict or drama.

Identify the themes in your story. Good winning over bad, standing up against bullying, or that trees are people too. Try to make sure these themes come out strongly in the novel and that every character and event is contributing to them in some way. If you want to add a chapter of silliness you can do so, as long as it is moving the book forward.

In the middle section: do you have the events you need to introduce the climax at the end? Have you connected your characters enough so their significance to the story is known (or is an intentional surprise)? List all possible endings. Which one is the most satisfying conclusion to your novel?

Reading 3

Time to get picky. Go through your manuscript with a highlighter and mark every adverb, every adjective and every needless word (i.e. 'actually', 'in fact', 'very', 'really', 'up' – as in 'stand up'). Look at each sentence, particularly the end. Is the strongest word at the end? Or have you ended every sentence with the word 'it'? Try to minimise the adverbs and look for more interesting words to replace them. Get rid of the repetitions (words or phrases). Have you used the words 'like', 'likely' and 'likelihood' all in the same paragraph? Change at least two of them. If you can identify the words you use all the time (I myself like the words 'tangle' and 'appeared' a little *too* much) search them out and destroy them!

Carefully look at how you start each sentence. Does every second sentence start with the word 'he', 'I' or 'she'? You need to vary how each sentence starts rather than start them in the same boring way.

Look for more precise words than the ones you initially used; is there another way to phrase what you are saying? Take out a thesaurus and make a list of every word that relates to your themes. Are you writing about loneliness, happiness, innocence, evil, the colour red? Look for places to insert them into the text so that your language contributes to the themes as much as the characters.

Be wary of sentences that repeat what you have said before. People will bear repetition, but only so far. If you are trying to reiterate the same thing more than once, three times is probably about as far as you can push it. When you are giving a speech you can do much more repetition because your body language, voice and pitch add something new to the message every time. However in a book a person can just re-read something important and they don't appreciate having to wade through the same stuff to get to something new. Reflect on what it is you really want to say and pick out the words that most simply and plainly state this.

A reader will appreciate your clarity and will be more inclined to re-read what you have written because it was written well.

Reading 4

Check for punctuation and **only** punctuation because by now you have the very best sentences in the very best order and can ignore those pesky words. Find the right place to put those colons and semi-colons we use to make smiley faces in emails ? ? And for goodness' sake, run spell checker! But remember, spell checker will not pick up everything, so a proof read is needed by a pair of human eyes too.

Final Pieces of Advice

- <u>Show Don't Tell:</u> if you have a paragraph describing what people are doing, ask yourself if it can be done better through dialogue. Rather than saying a character said something angrily, show it by having them stomp their foot or throw something at the cat.

- <u>The Hook:</u> make sure there is a moment of interest or suspense at the start and end of every chapter. These hooks will make your book the page-turner it is meant to be.

- <u>The Waffle:</u> many people like the sound of their own voice. Unfortunately, though we don't intend it, most writers love the look of their own prose. We tend to waffle. But to be a successful storyteller and conveyer of information we need to be concise. Can that two paragraph description be condensed into a beautifully worded three sentence description? Then do it and put the thought in to make it right. By editing out the rubbish we make the reader's experience more enjoyable and they will return to us. More is not necessarily better.

- Be Conscious: always be aware that you are crafting each sentence. All the well-known, highly regarded novels are there because the author thought carefully about how each sentence was worded and moved the story forward. Robert Kiyosaki's *Rich Dad, Poor Dad* would not have been nearly as famous if he hadn't carefully thought through each sentence. The same goes for J.K.Rowling, Stephen King, Lee Child and any other world famous author. You want to avoid clichés and be unique; this won't happen without constant awareness of your work.

2.3 Are You Living In A Little Pink Cocktail Umbrella Heaven? Doing Market Research

Have you ever entertained the fantasy of sitting on the beach, your laptop on your knees, a cocktail with a little pink umbrella in your hand and a crowd of excited fans around you requesting you sign their dog-eared copy of your novel? Have you even entertained the idea of living in a cardboard box under a bridge? We may as well be realistic here: the latter is the more probable, if less compelling, image of most of our literary prospects.

I know many of us picture the future of our writing career as a little pink cocktail umbrella heaven (myself included; fantasy writer = vivid imagination), but the cold, hard, mouldy cardboard smelling truth is not everyone is going to like what you write. Leaving your market research at, "Oh, everyone is going to love my book," is plain suicidal, with your self-indulgence and arrogance being the knife that makes the cut. We all need to find our selling point; another generic synopsis of a vampire novel isn't going to cut it anymore. Now if it's a space opera vampire novel you may stand a chance…. Let me explain.

First, you have to determine: what is your aim? In writing and publishing this novel, what do you hope to achieve? Do you feel compelled, feel you're talented, have an imagination that wants out, want to share your experiences, your opinions? If those are your reasons for writing you may find you audience more of a Shih Tzu size than a Great Dane size. My favourite reason is writing for fame and fortune.

Do me a favour and on a piece of paper write the name of every author you like to read. Now search online for a list of author names at a major writers' festival. Compare the number of people on the festival's program to the number of people you actually know the names of, and the people on your list. The number of authors you don't know far outweighs the number you do. Have all of them been published? Yes, several times generally. Are all of them famous? They'd like to think so.

If you are writing to make money then you had better make sure that you are writing in a genre or on a topic that people want to buy or need to buy. Just because you have a great idea does not mean that people are looking for it or feel that they want it or need it when they see it. I am an advocate of writing what you love and what you are passionate about. However, if you are writing to make money, you need to go about it differently. A great place to discover the things people are talking about is newspapers.

For the rest of you who have written what you love and have the philosophy that, while you can maximise your sales, you may not be the next J.K. Rowling, let's talk about market research. This is going to involve you asking a lot of questions.

Who will read your book? What age group? What gender? What interests? What side interests? What demographic (religion, single mothers, gay etc.)? You need to think about where your novel would be advertised/publicised and how you might reach that group. For example, if you write comic fantasy, your target readership may be young adults between 14 and 18, who like reading Terry Pratchett, school libraries and readers of certain kids' magazines, comics or video games. Then you need to find out how many of these readers/magazines there are.

Look online, type in key words (e.g. Terry Pratchett, fantasy, comedy etc.) and find out where these people congregate (look for online communities). Check if there are Google, Yahoo!, Meetup or Facebook groups set up by readers in your niche. Are there Facebook pages on how to eat chocolate and still not gain weight? How many people like these pages? Are there online forums on mountain bike riding down ski slopes? How many people have visited the site (see the visitor counter at the bottom)? What is the readership of the magazine *Soaps Not Gropes?*

Also look at your local government statistics websites. How many people attend university or school? How many declare Jedi as their official religion?

Find out how many searches are being done per month on your

topic, on a book similar to yours, in Google. There are several great tools to do this. One is Google's own keyword statistics tool which you can find in the Google Adwords site. This gives you an idea of how many people are searching for this particular keyword, and even suggests other similar keywords that people type in which you may not have thought of. It also shows how much competition you have for this keyword – basically how many people paying for advertising on Google are using this keyword. This will give you an indication of how many people are looking for what you are selling and how popular it is. The best part about knowing the potential size of your readership is you are under no illusions as to the size of your market when you set forth.

The great thing about Google Adwords is that even though you have to sign up for an account, you don't have to give your credit card information to do it, and you can access the Google Keyword Tool within Adwords for free. There are many YouTube videos showing you how to use the keyword tool (such as this one *https://youtu.be/ NfzyDg4XfL8*), the key thing to remember is helpful to determine the most searched words in Google for your topic/genre, if anyone is using the keywords and phrases you think they are, and what keywords and phrases are highly searched (and related to your book) that you *didn't* think of. This will then give you a great list to use later in any metadata you enter (more on that later) and any ads you might want to run. There are also other paid keyword trackers are available, like *www.wordtracker.com*.

Now that you have found these odd (yet wonderful) people, why would they want to buy your book? Is your book as good as, superior or similar to other books on the market? What can you do that is different or better? What are others NOT offering? Research similar books on Amazon, research their content, reviews and weaknesses (this is particularly relevant for non-fiction). How does your book provide what theirs does not? Is it easier to understand? Is it the first of its kind? Is there a unique theme?

For example, your vampire novel takes place in a space opera setting. Or you have a unique narrative voice, a different angle, or you cover something in more depth. There are millions of cook books out there but how many are specifically designed to use foods that help people with Alzheimer's? Find an angle that makes it unique. Please, whatever you do, do not promise that your book is unique and then not deliver. It's like taking a toy away from a sumo wrestler at Christmas – it's not advised.... or beneficial to your health. (Human pancake anyone?)

During your researching adventures you may come across a topic that is very similar to the one you are writing about, however it has thousands more enthusiasts than the market you are aiming at. You may want to consider tailoring your writing to fit what the people of this new market want. Can you include an extra chapter in your novel, or an extra character or concept that works well with your current work? If so, you may find that you can also capture readers from this similar yet more popular branch of your genre.

If you are an author who wishes to approach a publisher you also need to consider this question: why publish my book? What will convince them to spend thousands of dollars on it? There needs to be enough of a market to generate monetary return. Are there no similar books in the market? Does it deal with a current trend? Is there a similar book that has been favourably reviewed recently? You need to make your angle clear and concise. Searching your country's local government survey statistics website is also a great place to find information to convince a publisher of your readership's existence.

For example, for my gap year travel novel I found out from the Australian Bureau of Statistics that there are 425,000 students in years 11 and 12 and 56% of those go directly to university. After I had handed in my proposal to several publishers I found out that almost 40% of students drop out of their course within the first six months of uni. Now that would have really added to my case.

One final thing is to research where your competition is aiming their campaigns. What websites are they advertising on? How much are they paying for their ads and what is their unique selling point? This is one of the most powerful parts of this research.

Competition is fantastic; they have already set the stage for you. Then it's up to you to use your unique selling point to make their readers **your** readers. A fantastic site to stake out your competition for certain keywords is *www.spyfu.com*.

Now that you know who your readers are, where they congregate and what they like, now maybe, just maybe, you can allow yourself one cocktail with a strawberry on the rim at your local bar. You can work your way up to the umbrella.

2.4 The Foundation For Published Success: Pitches And Proposals

There is an excellent chance when you declare to someone that you are writing a book that the first thing out of their mouth will be the all-important question, "What's it about?" If you are serious about making a go of it, responding with an impression of an open-mouthed fish scratching its head is not the best way to face this question. You need a pocket sized pitch ready to go. When you approach an agent you have exactly 15 seconds to impress them with your pitch, NOT with a string of umms and ahhs. When your agent approaches an editor at a publisher, they will pitch your book to the editor, then that editor pitches your book to all the departments in the publishing house. This publishing sales force will approach the booksellers and the media with your pitch. These people do this for a living and they have the connections and the practice.

If you, on the other hand, self-publish your novel then your pitch has to be so good it will flatten the readers, the media and everyone else you have to deal with into cardboard cut-outs. If you're lucky, your pitch will last long after you're six-feet-under and providing food for flowers. A hundred years from now, when a reader is reading your novel off the inside of their sunglasses on their personal hover board and the person next to them says, "What's that about?" . . . you'd better hope that reader can give a sucker punch of a pitch!

It is essential to create three pitches so you have something hypnotic for every occasion. I'm talking seriously hypnotic, like getting an entire crowd to do the chicken dance kind of hypnotic. That is how much consideration you need to put into your proposal and pitch. If you cannot capture people in several seconds with few words, you're going to have about as much success as a marathon runner with two broken legs. These bite sized pieces of your work make a promise to readers: to educate, entertain, humour, inspire or scare the living daylights out of them. When you are writing a pitch and book proposal, on pain of bad sales you should avoid weak pronouns, adjectives, adverbs, jargon and clichés, and write in the third person regardless of whether your novel is written in first person, second person or iambic pentameter. The more professional you seem (it's all about the illusion here), the greater your chance of being read rather than being turned into a paper aeroplane.

The first pitch should describe your book concept in seven words or less. Basically, what type of book are you proposing? A biographical account of a turnip? A guide to making cheese omelettes? Adventures of a Ninja bowling league? Memoirs of a person who kind of smells? The pitch I have for my gap year travel book is "The ultimate guide to overseas gap years" or "Your international gap year made easy". Simple and concise.

The second pitch describes your book concept in 25 words or less. This answers the question, what is your book about? This needs to be simple, tight and catchy. It needs to highlight the benefit for the reader, the most interesting thing they are going to get out of the novel. The best pitches are those that play on the emotions and imaginings of an audience with phrases such as "for everyone who has dreamt of...". My pitch for my gap year guide went like this: "An info packed adventure guide to gap years for anyone who has dreamt of breaking free and seeing the world between school and university." The best ones make the benefit to the reader immediately apparent, as well as reading like headlines by being emotive and creating curiosity.

The third and final pitch is an expansion of your second pitch. It should be between 70-150 words. This pitch is your last chance to entice your reader to buy; waffle at your peril! It should summarise storylines and themes and keep the reader's attention until the last full stop. Here you need to make a big promise to your reader by outlining what benefits the reader will get for investing their time in your words. Use their imagination in relation to your book; make them wonder what their life will be like after they have read it. What will they see, hear and feel through experiencing your book? If the information you provide in your book is rare then tell them why. Frankly, there is only so much a potential reader can glean from that purple poodle on your front cover, so lead them to it.

3.4 Pitches And Proposals

These pitches will make the basis of your book proposal. If you believe that this is a little too hard for your tastes, you should first consider that not making any money because you were lazy is going to feel like a Taser to the private parts. If you want to sell an e-book or be offered a publishing contract then this is something you must care about. You have already spent years writing your novel; two weeks of inventive pitching is a little drop in the large ocean of creative chaos that is the life of a professional writer.

A book proposal should consist of a cover letter, author credentials, synopsis, market potential, chapter plan and sample chapters.

Cover Letter

A cover letter is more useful to those who plan to submit to a publisher or agent. Basically it includes a greeting, your second pitch to excite the publisher, a brief overview of why you are the best person to write this book and a quick overview of what is included in your proposal. However, it is possible for a self-published e-book author to use a letter as a way to address their readers. You could do this on the home page of your website for example. A little 'Dear friend' letter so to speak (e.g. Dear Friend, Do you have trouble x and y? Well you've come to the right place…). Author Jim Brown does this on his website to get readers to join his mailing list.

Author Credentials

Listing author credentials, or you could say experience, is just another way to convince the reader or publisher that you have done a good job of the book they are investing in. This is particularly relevant for non-fiction writers who generally have a body of experience that has prompted them to pen their novel. Do you have certain qualifications? Or have you just been to a few too many Mardi Gras? Whether it's experience, special insight, a similarity to your readers, a lifelong interest or writing awards you've won, listing these credentials is a great way to convince readers and publishers to trust you and your ability. If you are penning under a nom de plume (a pen name) perhaps use that mystery to boost both interest and credibility by explaining why you are anonymous.

Synopsis

A synopsis is not a laundry list of events that occur in your book. I don't want to see, "And then Benny went to war and shit happened," from any of you! Here you have to communicate who your characters are, their emotions and decisions they face. Here is where you make your genre clear and use your third pitch to excite the reader's imagination. If you write non-fiction you should also include a short bullet-point section with specifics on what will be covered and, more importantly, outline the benefits to readers in each of these sections. This should be no more than 200 words, for both fiction and non-fiction.

Market Potential

Knowing your market potential is vital for e-book authors and traditionally published authors. For an e-book author, a finely tuned idea of your market will make targeting readers and groups online faster, easier and ten times more effective. This will tell you what forums to join, what angle to write your synopsis from and what the most effective keywords are for you to use so readers can find you easily. After researching the competition you will be able to accurately tell you reader exactly why your book is unique and worth reading.

For an author approaching a publisher, having done market research shows how much thought you've put into your book and proves that a readership already exists for them to make money. Publishers, as mentioned previously, do not know every fan base or craze, but if you can show enough evidence of your potential readership's size and accessibility you will have a much greater chance of success. Here it is all about who, what, where and why.

Chapter Plans or Sample Chapters

Chapter plans and sample chapters give you an opportunity to show a publisher that you have done the research, organised it logically and creatively and have written an engaging and marketable book. A publisher will then have guidelines on their website specifying exactly how long they wish your sample to be. If they ask for three chapters, send them three reasonable chapters (amalgamating three chapters together so it looks like one is not a good idea…). Don't send them a paperweight; the bigger an unsolicited manuscript is, the less inclined they are to read it.

For an e-book author, a chapter plan helps you to decide what sections of your book are the most engaging and can be used as samples to draw the reader in. It's up to you how big these sample chapters are and we will discuss this in more detail later. Each chapter should have a compelling title (remember: abstract titles like "Bearded Mongoose" are not always the best descriptors) and several self-contained sentences that sum up the chapter.

If you aspire to e-book or publishing success these are the foundations for that success! Take the time to get it right, and you will be rewarded.

NOTE: This is just the tip of the iceberg, a summary of tips shared with *E-book Revolution Interview Series* (*http://www.cravenstories.com/books-and-more/courses/interview-series/*) listeners by expert Sheila Hollingworth.

2.5 It Started With 'The'

There is an awkwardness as you stand in front of the bookstore information clerk trying to describe a book you can't remember the name of. You think it started with 'The'. And the author's name was possibly Hobbs, but you're not quite sure. All you know is your friend described exactly what the book was about, in wonderfully vivid and exciting detail (using the author's carefully worded pitch....). It is a historical novel based in Elizabethan times as well as a thriller because someone is murdered, and it has an unlikely sci-fi twist that includes a UFO sighting. Or it's a book on fashion design specifically for making dog clothes. Or a book on how to make landscaping Feng Shui compliant. After searching through the computer the clerk finally shrugs and, with a very pointed look, stares over your shoulder at the next customer. As you shuffle out of the store, muttering insincere thanks, you decide, "Bugger it, I can't be bothered."

The same goes for searching, without help, through the online search engines. Except this method generally involves less awkwardness and more frustration. To the point where the searcher will not only decide to give that book a miss, they will actively hate it for wasting so much of their time. Making your novel easily found on the internet is more crucial then media attention – it is even more crucial than distribution. Because if you don't target your keywords to the right market you will be shooting blindly at people who don't give a brass wazoo about what you write. Get your keywords right and you get people like me, with the memory of a grandma with Alzheimer's, finding your book in the first search. You get it right and all those people looking for mountain biking, fantasy thrillers, sci-fi crime or a light hearted pick-me-up will be so excited they didn't have to waste time looking for what they wanted that they will snap your book up without a thought.

So before you go a step further, before you even let your tummy expand an inch forward, take out your proposal and your market research and jot down as many keywords as you can find. These are JUST WORDS. Perhaps three word phrases max. They should cover themes, genre, content, or the passions of your target audience. Basically what your novel is at its essence and what will bring people to it. Those free keyword tools will give you a fairly good idea of the terms that people search for to help you get the right words to describe your work. If you have at least 30,000 people per month (globally) searching for what you have, then you will more likely than not do very well. If the number of searches is less, you will need to be very specific about what words you use and where your groups interact online.

People search in keywords and they expect you to have categorised your work accordingly. People don't look for the words 'peril' or 'danger' or 'fantastic'. They look for topics and content and similarities. You can have the best synopsis in the word but it doesn't matter one bit if your keyword isn't typed into the search by a prospective reader. This is how you target your market and wave your red flag. This is how people in forums discuss your work.

This is also how the search engines like Amazon, the iPad store, Google and Yahoo! spit out their results. They don't wave a magic wand or partake in a bit of voodoo; they search methodically, like the machines they are. And for you to rank highly you need your major keywords everywhere! They need to be in the synopsis, in the title, in your advertising, in the tags on the e-book stores, in your website address, your blog name, your blog's website address, your Facebook page, your Twitter bio, your pitch, your forum posts, tattooed to your forehead, EVERYWHERE!

The more prominent they are the better chance you have of being found. Of course you have to balance being enticing and creative with the keywords. But if you make your book too hard to be found then that person who was looking for your Feng Shui garden advice is going to look somewhere else. Because frankly, they don't feel like looking the idiot in front of a pimpled, arrogant, teenage sales assistant.

2.6 Got It Covered?

Podcasts are a newly discovered passion. They are fantastic for making a work day full of spread sheets, models and tiresome co-workers go just that little bit faster. It was only a couple of days ago I came across the podcast of two book reviewers. They were discussing how the majority of e-books sitting in their email inbox had covers that looked like they were drawn by a child. And that child had probably been shotting tequila.

That age old saying 'don't judge a book by its cover' has been overused so much in society it has almost lost its meaning. But as a self-published author you shouldn't be so flippant. Whether you're trying to impress a reviewer, the media or your audience, you're cover is your best sales agent. It has to be all cheesy grins, slicked-back hair, flashy suits and a sultry voice. Because if you don't compel your audience, whatever their literary status, with your 'good looks' you get shuffled to the bottom of the pile and the next week you're eating stew out of a boot with a stiff shoe lace. Sometimes a front cover is your only chance to impress, and if it's not eye catching, if it's not intriguing, if it's not bright and jumping out of the page, a reader won't even go far enough to read those golden paragraphs you crafted for your synopsis. Their roving eye will pass you over as just another graphic. If you want to make money, you don't want to blend in with the digital slush pile.

For this you need a professional. Any successful self-published author, from the very famous examples like Matthew Reilly through to Amanda Hocking or Brian Pratt, will tell you exactly that. Image is everything. If you look like a bogan (red neck/hill billy/hick), talk like a bogan, drink like a bogan and pee on the side of a building like a bogan, you must be a bogan. Similarly, if you look like an amateur then your work is seen as amateur, inferior and not even worth a free download. Matthew Reilly, an action writer, self-published his first novel at 19.

He knew that though the editing, production and distribution was all his own work, if he got a friend or himself to design the front cover he was royally screwed.

There are many places you can go to get a professional front cover such as online creative websites like *www.upwork.com* or *www.99designs.com* where artists try to out-bid each other to do your work. There you can hire anything from a designer, to a ghost writer, to a levitating website guru (well, maybe not that far). Another great way to find designers you like is via the Ebook Design Awards run by Joel Friedlander of The Book Designer, who runs the competition every month. It was how I found the designer I regularly work with, Kit Foster.

Remember, in the digital world you don't have a book spine and you don't have a back cover; the front cover is your only visual selling point. Make sure it is eye-catching, unique, bright or has a good contrast in colours or shades. Any rule of photographic composition can and should apply to its design. The book also needs to convey the themes of the novel. It needs to convey drama, romance, stupidity, humour (depending on your content). See if the font used for the title can add to the theme and demeanour of the novel. The more information you supply to your designer the better. Give them your synopsis, your pitches, your keywords and any strong images you have of different tense scenes or landscapes. You also need to be strong about your vision; if what the designer has produced does not sit right then you need to discuss it with them. Why did they put that element there? Could they make this darker? Print out some other covers in your genre, line them up next to your book cover and see if yours stands out.

So it's time to decide: are you going to have the cover designed by an inebriated child, or are you going to have a cover that flashes a million dollar smile and gives you a brand new car?

To get the nitty-gritty of e-book cover design, I recommend download the E-Book Revolution Podcast Interview I did with my cover designer Kit Foster at: *http://www.cravenstories.com/2012/10/23/ e-book-revolution-episode-2-great-e-book-cover-design-with-kit-foster/*

2.7 Formatting, Decrease The Number Of Grey Hairs

It may come as a surprise to you that e-books are not like print books. I'm not talking about being able to sniff, fondle, bend and lick one and not the other. Please, get your mind out of the gutter. No, I am talking about formatting. Unlike a normal book, where everything is in its final resting place, a conventional page just does not exist in an e-book. Readers can and will manipulate the text to suit them, to make it larger, smaller, to spread apart the lines, or adjust the font. What we as authors need to realise is that an e-book needs to be a continuous flow of story and trying to make the formatting conform to that of a printed book is going to get your knickers in a twist quicker than you can throw a computer out a window.

The main platforms for pushing our work out into the great unknown are Amazon, Kobo, iTunes, Google Books, and distributors like D2D or Smashwords (which distributes to Barnes and Noble, Apple if you don't have a mac to upload separately, Sony, Diesel etc.). Each have a slightly different way you must format your book before it is uploaded into their system. Formatting is about making the reader's experience more enjoyable, so by the time they have finished our masterpiece they are in such a state of ecstasy that they are screaming for your next novel. To make your formatting the best it can be, Smashwords has a Style Guide available for free on its website (*http://tiny.cc/swstyle*) and Amazon Kindle has a help section devoted entirely to formatting, which you can access at *http://tiny.cc/astyle*. Other sites require you to already have an epub file handy for upload or a .doc or .docx Word Document such as Kobo or Apple iBooks.

Though, a warning, the conversion from Word documents is rarely perfect, so having a program that can pre-make an epub file is normally best. Free programs like Calibre can convert Word documents to epub files with a bit of practice, as can paid programs like Scrivener.

However, I am a firm believer in being prepared so that when I get to the nitty gritty stuff I don't have to spend what little free time I have shouting at the computer screen. There are less police interventions that way. So, having waded through the various publications on ebook formatting, I have pulled out some tips to keep in mind so your trip from creativity to saleability is that much faster. Remember: not following the tips below will usually result in your file not being accepted by the various platforms.

Top Tips For Less Grey Hairs

- You will make it 10 million times faster for yourself if you type your creative prose in Microsoft (MS) Word from the get go. Smashwords only accepts MS Word docs (they must be saved as a .doc file –'97-2003 word docs – NOT a .docx). Otherwise you must convert your novel into an MS Word file before you begin. Smashwords (SW) does take one or two other types but it says Microsoft is the preferred way. So stop making things hard for yourself and just do it!

- Make a copy of your document first!! Do not make changes to your original document; if something goes pear-shaped you want to have your original on hand.

- DO NOT use the space bar or tabs for indenting the beginning of your paragraphs. Use the indent function in MS Word (see SW style guide).

- Don't hit the enter (or return) key more than four times. This creates blank pages in the e-book and will cause an error in the conversion system when you attempt to upload your work. We want to fill our books with words, not air!

- Choose to either indent the start of your paragraphs or use block paragraphs. It's one or the other, not both. If you choose to neither indent or use block paragraphs, all of your paragraphs will run together. A reader only needs to take one

2.7 Formatting, Decrease The Number Of Grey Hairs

look at that and they will be running for the hills. Kindle also specifies one or the other. Within Kindle compatible products, the paragraphs indent automatically. However, if you want to specify how far they indent you can apply the formatting language as shown on the help board.

- Don't use fancy, non-standard fonts (for best results use Times New Roman or Arial) and keep font sizing at no more than 16pt in size. The basic rule here is to keep it simple! Keep bold and italic text to a minimum. Otherwise you may find you do a bit more fiddling than you bargained for.

- Columns and tables are not supported. If you really need tables put them in as an image.

- Wrapping text around your images is a big no-no for most major e-book sellers. Have the images on their own line and centred in the middle of the pages. It is very hard to attach words to a specific image so either make the words part of the image or get ready for some serious, frustrated hair pulling...

- Most platforms will limit the file size. So if you have several images it would be wise to use the compression feature in MS Word to shrink the size of the document. Don't worry; it won't affect the quality of your images significantly.

- If you only have a print book copy of your novel there is still hope! You can get the novel scanned by various companies and have it sent back to you as an MS Word document. They are generally very accurate but you should still carefully proofread the document before you begin to format it.

- In Smashwords, if you want a front cover to appear in all formats you need to have it as an image on the first page of your MS Word document. Only certain types of formats produced by Smashwords allow you to add the cover separately.

- For all you budding poets: you need your poetry to either be left-justified or centred, otherwise who knows what order your words will be in.

- Amazon asks that you use page breaks to separate your chapters. These can be inserted in MS Word. However in most formats in Smashwords these page breaks will not be converted, so it's best to put one or two enters (returns) above and below the page breaks if you decide to insert them.

- When noting your copyright use the word rather than the copyright symbol because it doesn't convert well.

- You need an ISBN number to distribute to Apple and Sony.

- You need to include front matter in your book, such as a centred title page including your carefully thought out and intriguing title and your name (or nom de plume if you prefer). It is also recommended that you put a copyright statement in. Make sure this page is your most professional! It is the reader's first impression so, unless you are a HTML guru, skip the fancy stuff. It's also recommended by Smashwords (who publishes DRM – digital rights management – free files) that you give a gentle reminder to customers to refrain from piracy. There are examples of this statement in the style guide.

- If you're keen on adding a little colour to your novel you can add a simple glyph between paragraphs to indicate a change in scene.

- For the non-fiction authors: you can also add a linked table of contents that allows a reader to jump straight to the chapter they wish to read.

- To upload to Kindle you need to save your file as a filtered HTML file in MS Word as per the Kindle board instructions. You then upload this into MobiPocket Creator where you

2.7 Formatting, Decrease The Number Of Grey Hairs

will add your front cover and convert your e-book to Kindle format.

- Once your file has been accepted and converted you must check your work!!! Make sure it translates well in each format and doesn't cause the reader to cross their eyes and fall sideways off their chair.

Have fun!

2.8 Avoid the Meat-grinder! Creating an Epub File can be as Easy as Writing a Blog

I always had it in my head that Smashwords would be my first e-book platform, the easiest, the largest distributer, the least work. Then I looked at the style guide, and read the tax information and realised I needed to buckle on my superhero suit for that gauntlet. I'd had enough trouble deciding how to format my YA novel to look like Facebook, now I had to figure out how to stop the meat-grinder from turning my work into something resembling a 52,000 word experimental poem. It was still on the cards, but I needed a new plan to keep the momentum rolling before I drowned in paperwork and coding.

I have always been a keen advocate for self-education. In industries like writing and in particular e-publishing and internet marketing, things change too quickly for your run of the mill University. It's this drive for self-education that has seen me being asked to present on marketing and creative e-book advances, and got me into the coolest courses for free. It was at one of these courses I found a way you can create your own clean epub file, for free, without any supernatural talent at html code (Or Smashwords formatting), and have your books uploaded on Amazon KDP (Kindle Direct Publishing) and Kobo Writing Life in the same hour.

PressBooks is God's gift to writers.

PressBooks (*www.pressbooks.com*) has taken the WordPress open source code and created a powerful tool to produce epub and print-ready PDF files. The epub file it produces is directly uploadable to Kobo Writing Life and KDP (Who convert the file cleanly to mobi) and it's as easy as using the WordPress blogging platform.

There are several reasons why I would marry PressBooks over my partner of five years:

- The platform is free to use. You can create a file for every one of your 100 haikus if you wish.

- It is as idiot proof as blogging with WordPress. Basically you set up each of your chapters as 'individual blogpost'. This makes it incredibly easy to fix your own typos and add extra scenes or features you thought up while singing in the shower. You can upload and insert images in the same way you would a blog, easily compressing or enlarging your author photo depending on the size of your narcissistic streak. In fact, I'm using it right now to update this book.

- It allows you to place your front matter (copyright message, dedication, foreword), Novel (separated into chapters and, if necessary, Parts) and back matter (about the author, more books from the author and bonuses) on separate pages avoiding the run-on effect between sections that you get through the Smashwords conversion. You can select which components you want to include in the epub export with just one click, allowing you to create unique versions for different markets (Kobo Users, Kindle Users) or price points (e.g. I have a multimedia version of E-Book Revolution which includes links to ten instructional audios and a private Facebook group. This version of the e-book I sell for $19, where as the e-book without audios is only $2.99).

- It provides three different template styles, all super professional. Or it allows those of you who have painstaking learnt html code (who I like to call the 'obsessed' or 'book designers'), to upload your own CSS style sheet.

- It allows multiple users to work on a novel at once. So if you are a small publishing house you can have the author upload their work, then a structural editor, a copy editor, a cover designer, and a copywriter all do their work on the same platform. PressBooks allows you to compare the changes between saved versions so the author/editor can approve

3.8 Creating an Epub File can be as Easy as Writing a Blog

changes. Each individual contributor can access the novel online from wherever they are. The efficient work flow should have all publishers salivating, imagine the best of the best being able to work on the same project and not even have to be in the same country.

- Finally, but most importantly, PressBooks gives you the ability to integrate detailed metadata into your book. For those of you who have cocked your head to the side like a rabbit in headlights, metadata increases the find-ability of your books in the search engines ten-fold. You can add keywords or tags, ISBN's, pricing, synopsis, subtitles, covers etc into the metadata and it will be integrated into the backend of the epub file, just waiting for Kobo, Google or Amazon to search it.

Creating an epub use to be a mountain (or a giant hole in the wallet), now the only (mother of) a mountain is marketing and the reader author connection. That's where section four comes in… Sigh. Has anyone seen my Superhero Suit?

2.9 How Non-US Authors Can Avoid Paying 30% Tax To Smashwords and Amazon

It's sad but true that the only time you can avoid tax is when you're dead or homeless, and let's face it, I'm not that desperate just yet. Even in the online realms it's unavoidable, governments are the biggest sticky beaks particularly when it comes to money that is not theirs. Another unfortunate reality is that though you may be able to make your book available worldwide, the services you are using to do this still have their roots firmly planted on US soil which means the US government have a claim on your earnings, even if you don't live there, even if you are not a citizen.

If you self-publish on Amazon, Create Space (Amazon's Print on Demand company), or Smashwords, they will withhold *30% tax* on your royalties unless you have a US tax ID. If you are a non-US author (in particular Australian or NZ based) and you DO have a US tax ID you only have to pay 5%, so you end up with a whopping 25% more royalties by doing something about it. I was pretty excited by this, then I looked at some of the instructions on the internet and realised that getting a US Tax ID for an individual is a pain in the butt.

Well, 6 years ago it was. Back them you used to have to obtain either an ITIN, EIN or SSN from the IRS which involved calling the IRS during opening hours, usually on Skype because it was cheaper, and then fight with the operator to convince them that yes you could apply for said number and these were you details. Then you filled out a paper version of the form and snail-mailed it. Yawn!

These days, all you need is the tax number you use in your country, in Australia for example, as soon as you start working you apply for something called a Tax File Number. This is what you would use to fill out the IRS tax form known as a W8-BEN for each distributor (Amazon, Smashwords, Apple iTunes etc). Below are some simple instructions to help you in your mission!

- You only need to complete a W8-BEN form for companies based in the US (for example, Kobo is based in Canada and doesn't require one).

- For Amazon Kindle and Smashwords, this form can be filled out online in your dashboards. Though if you'd like to download a PDF version of the form or need instructions on how to fill it out I have links for both of these in the important links chapter of this section.
- In Amazon Kindle, you will find your Tax form under Your Account – Tax Information. Then select View/Provide Tax Information

- In Smashwords you will find your Tax form under Account – Payment Settings, then scroll to the bottom of the page to view your tax information

- For AUSTRALIANS: The tax treaty between the US and Australia is **Article 12.** This allows you to pay only 5% withholding tax on book royalties. For OTHERS: you can find out your tax treaty number in this document: *http://tiny.cc/taxtreaty,* but many of the online forms will find the correct article for you based on your country of residence.

- For the reason you are using the W-8BEN form you put "Australian (or UK/Swedish etc) author self-publishing via US distributor"
- For Amazon make sure you use one of these titles next to your signature: CEO, CFO, Treasurer, Company Secretary, President, Vice-President, Director (UK only), Managing Director, Executor, Member, Managing Member, Partner, Managing Partner, Self (which I use, if you are an individual author use this too) and Trustee.

- For Print on Demand accounts like Amazon CreateSpace follow the same instructions for the Amazon form.

3.9 How Non-US Authors Can Avoid Paying 30% Tax

All I can say is, you lucky butterflies, you didn't need to go crawling through the mayhem of 2012 to make this tax break happen! Hopefully this run through has put your mind at ease.

2.10 The Superbook

The word 'test' inspires horror in many people and may even cause some to break down in a quiet corner of the room. But the word 'test' is a marketer's best friend and, whether you like it or not, the time has come when you need to swap your creative brain for your analytical one. Because if you want to make money and do this for a living then every little detail must be honed to absolute perfection so that you have a well-oiled, reader catching machine.

In some ways this requires you to be extra creative, because in this step I want you to do at least two of everything. That's right, two; I want your creative mind to be so drunk it's seeing double, triple, quadruple! The drunker it is the better. You want to make sure that the synopsis you have, the headlines you use to advertise, and the title of your book are going to entice the greatest number of readers. However, if you only have one synopsis, one headline, one price and one title you will never know that if you had just changed the word 'sultry' to 'seductive' you would have sold an extra 10,000 copies. This technique is known in the marking world as split testing, and no, they are not practising gymnastics – it is how they make their money.

So how could split testing be applied to an e-book? In an ideal world the publishing platforms would allow you to upload two different versions of your titles, your blurbs, your keywords, your covers, your pricing, and then they would randomly show readers the different versions. But this smart business strategy doesn't seem to have made it into the publishing tools yet, so you will have to do it the old fashion way. Change one thing at a time, and give each change at least two weeks to prove it's worth (or one week if you're a conscientious marketer). I would not recommend having any more than three different versions of each thing you wish to test, otherwise it's going to take you *years* to fine tune things. In saying that, test as many things as you can *before* you publish too, such as titles for your book, or cover art. If you already have friends on social media, a blog or a writers' group, come up with five or six headings, and maybe two or three covers, and then find out which one appeals to your test dummies more. Then use the top picks for your initial novel upload. In a month, the real testing to find your superbook combination can begin!

One rule that you must follow before all others is to only ever change one thing at a time. If you change more than one you are going to have a hard time figuring out which one made your new readers download happy. It's similar to only wearing one pair of underwear a day; if you change it three or four times you're going to have a large laundry hamper to sort through. For each new change compare your previous weeks' worth of sales to your new sales. Have the changes had any effect on the sales of your book? If it's for the worst, revert back, or try a new test. If it's for the best, move onward with another element such as a book blurb or pricing.

Next determine whether giving your novel away gets you more readers than giving away a 100 page sample of your work. Do the same thing; test it for a couple of months to see what happens to each. Then test the pricing: how many people buy it at 99 cents in comparison to $1.99? In comparison to $2.99? Remember to only change the pricing a little at a time. Are there less people buying the book at $2.99 but you are getting more money? There are some fantastic discussions on J.A. Konrath's blog about pricing (*http://jakonrath.blogspot.com*) so have a look through those and decide on a step-by-step testing process for your two novel versions.

Next have three or four different versions of your synopsis. I would do this after you have decided on an optimum price. How do the first two compare? Take the one that performs the least and replace with another synopsis to test against the stronger one and so on.

If you are advertising, look at changing the wording of your ads to see if one ad brings more traffic to your page than others. Always refine your keywords by checking what words people are using in their forum posts, their Tweets, in hashtags on Instagram, and on Facebook, and keep adding to your keyword list (your 'tags') all the time. Test your cover art as well.

The best way for this testing to succeed is to make sure that your novel is available on as many platforms as possible. The more places you are available, the more chance you have of being seen.

2.11 If You Want To Be A Great Writer, Educate Yourself

I started writing my first novel when I was thirteen, inspired by my writing idol Isobelle Carmody who had started her first (eventually published) novel when she was fourteen. "If she can do it, I can do it," I reasoned, in fact I told myself I was clearly better having started younger. What can I say; I was a precocious kid and probably deserved a few mud pies in the face. I happily sent the finished novel off to half a dozen publishers at the end of high school, exceptionally proud and quite certain I *was a literary God*. Well, before that first round of polite rejections rolled in. Then I got 'The Talk' from my parents about making sure I had a good job before I started playing around with 'all this writer stuff'. 'You can do it in your spare time as a hobby', they argued.

When I finally finished University and got a job, I realised Plan B sucked a whole bowl full of rotten tomatoes. People did things they hated for twenty years?! Were we a whole society of nutters? Why can't I just repeatedly hit my head against a wall and step in front of a bus? I'm sure that would be less painful than data entry.

Because I had done so well in University my early egomaniac self still had the power, it still believed getting into the writing game on mere talent alone was a given. Don't pretend you haven't been there, that because you can read, because you've been writing since you could hold a crayon, because you can text your mate, or got a great response to that post on Facebook, you believed that writing a fantastic story would be easy. It was all about talent, you didn't need to build on your skills, you knew what worked. Needless to say 'Being A Writer Take Two' burst my ego enlarged head with a nail gun. The story and scholarship application rejections slapped me upside the head instantly and if one kind stranger on a scholarship board had not made the below suggestion, I would have called myself a talentless squirrel and given up life for a hippy commune:

"You're obviously very hardworking, but just don't have the experience in the publishing industry. Join your local writers centre, be a part of the local writing community."

Yep, I was a dumbass. Why I thought the creative arts didn't require lessons I don't know, but it made a lot of sense. So I hired an editor, I took writing courses at my local writers' centre, I actually got feedback from other writers and learnt that my writing sucked. Big time. I was so embarrassed at what I had sent to publishers I could have powered a small city for a month on my shame alone. I worked hard to improve and hone my craft, and that work paid off when I landed a mentorship with one of Australia's biggest fantasy writers (Isobelle Carmody, my idol), a job at one of Australia's largest writers centres and national and international speaking gigs. My latest milestone was being asked to do a public reading of my work.

So this is my trick to getting ahead in your writing:

Writers write, but great writers educate themselves.

I can say without a doubt that the only reason I have come so far in my writing career is because I've constantly educated myself these past fyears – and I've been writing since I was thirteen. Every year I have a budget to spend on learning new things. Frankly, it was the best investment in myself I ever made.

So here is what you need to do to reach *your* full potential:

Talent is Not Enough

Yes you can spell and join words together in a sentence. That sentence may even make sense to people in various countries. You may even be able to put 300 pages of sentences together. But are those 300 pages worth people paying money? If you're a crime writer, does your book have a John Doe within the first scene? Did you even know you needed one? Have you put your red herring in the right place? In fact, have you even put your punctuation in the, right! place?

Seek Out Experts

I really resisted having to pay someone to improve my writing skills and industry knowledge. What did these people think I was? A money tree? I'm trying to make a living here! Then I compared how far I had gotten in six months doing my own 'research' in comparison to eight weeks taking a social media course and realised paying for information got me better quality, and I got to where I wanted to be faster. No one else was going to invest in me and my books if I didn't invest in myself.

While free information was great, it wasn't particularly detailed or step-by-step, no one was helping me step back from my own manuscript and point out what I couldn't see in my own work. I needed people who had more experience than me, and those people, the people that would really make a difference to my writing, wanted to be paid for their time. Experts hold back the valuable part of their knowledge (Usually the 'how do you do it' or 'how do you apply it to you' part) because they *know* its valuable, I could search as much as I wanted on the net but I would only get an idea of what I needed, not *how* to do it.

One Teacher Does N*ot* K*n*ow All

Do I still invest in courses even though I am now asked to give my own? Of course, I don't know everything. New crazes and ideas and strategies come out every day, and I try to share as much of what I absorb as I can on my blogs in the hopes that you will find it interesting enough to learn more, from me or someone else. I now know that I will always need an editor to look over my books, that there will always be someone I can learn from who knows more than me. Different writers have different tools, and some of those tools are going to make me want to gouge my eyes out and eat them, and others are going to work for me. I'm always seeking another way to jump start my creativity, skirt writers block, or create a unique character that scares children at night.

Set Yourself A Budget & Assess What You Don't Know

Nothing becomes a reality unless you make a plan and act on it. So start saving for a yearly education budget (hey, if you make money as a writer, it's tax deductable!) and discovering your weaknesses. Apart from improving your own skills, this also helps you link with writers who have better connections and can introduce those connections to you.

You Need To Learn About More Than Writing

Unfortunately the days of letting your publisher do everything for you are dead. Even with traditional publishers you need to know how to market yourself, and you may even find that having a strong online presence is what gets you that contract in the first place. The publishing industry is going through its greatest upheaval in 200 years and you need to know where it was, and what new opportunities are now out there. It's not all doom and gloom! Just another way of doing things that gives authors more freedom.

Pass On Your Knowledge

There is this myth that writers are all in competition. But there are only so many books you can write in a year, and once people are done with your books they're not going to twiddle their thumbs waiting for the next one. If it weren't for the numerous bestselling authors who gave me the time of day I don't think I would have gotten nearly as far as I am now. Have time for the newbies, they've got a dream too. You were there, you were probably more annoying...

So get rid of the egomaniac and get on with things. It's time to start treating your writing with the same seriousness as you would any University degree. Then one day I want to see you, passing on your knowledge to the newbies, offering a helping hand to others in the same way you were offered a hand by authors above you. You are better than your Plan B!

Umbrella 2: Important Chapter Links

Audio

- Audio 3.2 There, Their or They're? Self-Editing: *http://tiny.cc/audio3pt2*

Podcasts

- Cover design with Kit Foster: *http://www.cravenstories.com/2012/10/23/e-book-revolution-episode-2-great-e-book-cover-design-with-kit-foster/*

Video Tutorials

- Research your market (Google Adwords Keywords Tool): *https://youtu.be/NfzyDg4XfL8*

Tax Exemption – Non-US Authors

- W8-BEN form PDF download: *http://tiny.cc/W8BEN*
- Instructions on how to fill out the W8-BEN form: *http://tiny.cc/W8BENinstruct*
- To find tax treaty for your country: *http://tiny.cc/taxtreaty*

Finding your audience

- Yahoo groups: *http://groups.yahoo.com*
- Google groups: *http://groups.google.com*
- Meetup: *https://www.meetup.com/*
- Google keyword tool: *http://tiny.cc/Googlekeyword*
- Keyword tool: *www.wordtracker.com*
- Keywords the competition is using: *www.spyfu.com*

Other Links

- Online editing community: *www.fanstory.com*

- Online editing community/publisher: https://www.wattpad.com
- Online editing community: http://www.scribophile.com or www.critiquecircle.com or https://www.reddit.com/r/writing/
- E-book Revolution Interview Series: http://www.cravenstories.com/books-and-more/courses/interview-series/
- Freelance website/cover design: https://www.upwork.com/
- Freelance website/cover design: www.99designs.com
- Ebook Cover Design Awards: https://www.thebookdesigner.com/2011/08/monthly-e-book-cover-design-awards/
- Smashwords style guide: http://tiny.cc/swstyle
- Amazon Kindle KDP style guide: http://tiny.cc/astyle
- Simple e-book creation tool (this book was made with this): www.pressbooks.com
- Ebook Conversion Software (Calibre): https://calibre-ebook.com/
- Writing Software (with build-in epub file creation – Scrivener): https://www.literatureandlatte.com/scrivener/overview
- Pricing discussions – J.A. Konrath's blog: http://jakonrath.blogspot.com/

Umbrella 3: The Hype – Marketing

3.1 Marketing Starts Yesterday

Download the audio of this section at: *http://tiny.cc/audio4pt1*

I'm going to mention that horrible word 'free' and 'your novel' in the same sentence. Yes, I am completely serious. No, I am not in a straight jacket. No, I do not feel the need to jump off a bridge. There is a difference between those who succeed with their e-books and those who don't. Those who succeed are the ones who have learnt how to use free to their advantage; they are the Masters of Free and have made thousands.

Many interesting marketing skills and strategies (including viral marketing) can be very effectively applied to publishing, particularly in terms of niche markets. It is up to authors to market their books smartly. The author is competing for the reader's attention. Don't waste the opportunity; capitalise on the 2.5 million searches (per month) for free e-books on Google! By using techniques such as giving away the first book in a series or allowing a 100 page free sample for a 300 page book, you can hook the reader's interest and reel them in to buying the rest of the book/series. Free eliminates the risk for first time readers and builds you a following. If these readers love the book they will tell more people and help you to overcome obscurity. The lower the price of the novel, the more likely people are to buy it, increasing the number of readers and overall profit. This will not be the first time I say this: readers are an author's sales force!

There goes that angry 'teaspoon in a blender' noise again: "If I'm giving away my work for free, how the hell will I make money? Free doesn't pay the bills! Free does run my internet! Free doesn't let me flip off my boss and shout, 'I QUIT'!"

What you need is a following, and if you do not have a following then no-one will know who you are and they will not look for you or your books. You remain a nothing author. There are 2.5 million searches every month for free e-books; if half of those are just the same people typing the same keywords that is still 1.25 million people looking for a free e-book. By making your book available for free for a month or two, or adding free bonuses with your book, such as free audio, you are getting people to sample your work for no risk. And hey, at least *someone* is reading the damn thing rather than it just sitting in your desk draw or in an envelope in the middle of some publisher's slush pile. You need people to talk about your work and to anticipate it.

One great way to convince others to buy it is to give it away for free on the condition that they give you a review of your book. If even 10% of them remember you still gain testimonials to convince others to buy it. If you then sell your book cheaply or sell others in your series cheaply, the testimonials and your low price will get you more sales and gradually you can put the price up. Another great concept is to give your book away for free in exchange for the reader's email address via websites like Bookfunnel or Instafreebie.

With a hardcopy we cannot change the cost; it's one price and if the person doesn't like it then they won't buy it, no matter how pretty the cover looks. Digitally or in print, you have to get a following and reviews to sell. It just happens that online this is easier to do because, let's face it, you can't give a physical book away for free – it costs too much money to print, and unless you have bottomless coffers, your partner is probably going to confiscate your stash until you behave. However, you can give away a digital copy for free and you don't have to pay to print it or mail it to the reader. In the online bookstores a 15% commission on $0 is still zero, no matter how you look at it. If you have given away something for free people are more inclined to respond in kind with a testimonial or review.

There are many authors who argue that pricing a book low gives the readers the impression your work is amateur. I would argue that there are ways you can minimise that idea of amateur quality by making sure your cover is professional and eye-catching and that your synopsis is some of the best writing you have ever produced. You can use online writing and editing communities to hone your synopsis along with your novel samples. By making your work appear professional, you turn that 'low quality' impression into the reader feeling immense luck at finding such an appealing book at an affordable price.

The most successful e-book authors will tell you that the next step once you have written a great book is to write another great book to start building up a backlist, because each new book offers an opportunity to sell the others you have written. In this way you can make the first book free and be confident that at least one in three will buy the rest for $2.99/$3.99/$5.95 or whatever you decide. As long as you make an impact with your first, readers will want to know what happens next.

If there is a strong response to your e-book you can start looking into print on demand (POD) hardcopies. Cory Doctorow maximises his following and profits by simultaneously releasing a free digital version of his book and his POD at the same time. It is important to note that the attention span of a reader and searcher extends to a maximum of about five minutes. We're busy people, so if you don't have what we need right now, well too bad, we'll just go back and buy that book about flying pigs we saw. Once you know you have a popular book, by releasing POD at the same time you can consummate the sale quietly and efficiently before you've lost their attention and access to their buying impulse.

It is wise to have patience when first starting out. Most print book authors have taken years to get to where they are now, and the same is true of e-book authors. Writing is a long term investment, and e-book success is a slow build because you need to assemble a social network and create trust with your readers. And above all, marketing starts yesterday.

3.2 Websites: How To Entice Readers Into A Buying Extravaganza

One of your main missions as a successful writer is to be accessible to your readers and to create a community where they can interact with you and your work. So your mission, should you choose to accept it, is to give yourself a permanent web presence. It's time to delve into creating your own website. Here the advice 'the simpler, the better' rings loud and true. At least until you have enough money to make it flashy! So today we are going to talk about not only how to physically set up a site but also what you MUST put on there to draw your readers in.

There are three main avenues to fashion your own website. You can create a professional one with a dot com web address yourself, you can hire someone to do it for you, or you can use the free power of blogging. If you wish to create a simple website yourself, but you do not know how to work that Microsoft Front Page thingy, do not despair! There is hope. Several simple tools are available online, one of which is *www.squarespace.com,* or *www.wordpress.com*. Squarespace and WordPress will allow you to easily and simply put a website together and it will host your site for you. There are more tutorials than you can poke a stick at on how the programs work, so there is no need to stare at the computer screen in a hopeless daze. Nor is there much chance of your head imploding from staring at HTML code for a minute longer than you should.

If this is your chosen path, young hero, you shall need to purchase a domain name (address) from a website like *www.bluehost.com, www.hostgator.com* (for domain names see 'Our Services' at the bottom of the page), *www.godaddy.com* or *www.namecheap.com*. You can purchase the rights for a name of your choosing for up to ten years (depending on how confident you are in your quest or how many gold pieces you have in your pouch). Squarespace and WordPress allow you to transfer the domain name to their hosting service so you can then design your portal (website). It costs around $20-$30/month for these sites to host your website. Basically you are renting space on a server (known as a host) so that people can access your site. Host Gator has the cheapest hosting available for around $3-$5 a month for unlimited space and bandwidth, and they have a free site builder as well. This site builder is a little harder to use but for the money saved ($15/month) it might be worth an extra couple of hours. You must have a domain name and host before you can achieve your quest (for your website to work)!

Alternately, you can hire a trained assassin (professional) to build your portal (website). Upwork is a fantastic place to outsource (*www.upwork.com*) as is Freelancer (*www.freelancer.com*). You set up a description of what you want your website to do, how you want it to look, how many pages you would like and so on. Then website designers bid for the chance to work with you. So you can choose an assassin (website designer) based on their price, the experience they have, the interests they have, the websites they have done previously and previous customer testimonials, so you get the best person for the kill, erhmm I mean job. In this instance you can use a site like Host Gator to buy the domain name and host the site for as little as $5 a month and then just hand it over to the professionals. Remember to always research the previous jobs and testimonials for the outsourcer to make sure you get the most dedicated person for the task.

Or you can take the free road (and the long website extension) of setting up your own blog. Blogs are great for two reasons: firstly, they are free; secondly, the more you post on them the more visible you become on Google. Google likes sites that continually add fresh content, and if you start getting traffic (people) to your website, Google likes you even more. The biggest free blogging website is WordPress (*www.wordpress.com*). I should explain there is a free blogging version of WordPress and a paid version (which has hosting, domain use and better templates included). The difference is in the free version the domain name is [the name you pick].wordpress.com. Where as the paid version allows you to pick your own domain name (such as my current one, www.cravenstories.com).

Right, so you have chosen the path you will take on your quest

4.2 Websites: Entice Readers Into Buying

(whether it be self-made website, professional website, or blog) but the question remains: what needs to be present on this website to weave a spell around our potential readers and draw them carefully into our web? What we want to do is entice readers to buy. The best way to do this is to provoke an emotional reaction, because if someone feels emotionally connected to your work and can imagine how it relates to their own life and enjoyment they will buy your novel above the masses. You just have to make that connection.

There are several great steps you can take to increase your connection with your visitors and in turn make more sales. Of course here is where your synopsis will come out again, that brilliant piece of writing chock full of all those keywords you have researched. But you want to add to that experience both visually and audibly. So when a person visits your site you want to greet them with a video or audio clip. You can record a video with the webcam on your computer or with your digital camera then upload it onto YouTube. There are tutorials on YouTube that show you how to put your video onto your blog or website. Trust me when I say that barely anyone is doing this at the moment, but it is so powerful. People see you and hear your voice and they connect with you on a deeper level than they would through words on a page. In the audio/visual and digital age, this puts you a step ahead of the rest, and every step counts when you are self-published. If you believe you look like the troll under the Fairytale Bridge you can just record your voice and similarly find YouTube videos telling you how to upload your audio to your website.

The next thing readers should see on your website is your headline. Not "This is Bob's Blog" or "Welcome to the world of FrankenLou." The headline needs to grab attention! It needs to offer a big promise to the reader. Technically, what it needs to contain is a benefit to the reader for staying and an element of curiosity. This is probably easier for non-fiction than fiction. For example, an appropriate subtitle for my gap year travel novel would be, "Learn how easy it is to have the travel adventure of your dreams… with almost no money up front!" The adventure of my dreams is easy? And I need almost no money?? Please tell me what else! For fiction you want to appeal to an emotion within the prospective reader by relating the story to their needs and interests then, as with non-fiction, end with a curiosity. What we are aiming to do is excite the imagination and use the reader's imagination in relation to our book. They must see, hear and feel what it is they do or don't want, and wonder about how their life will be changed (I hate to use a cliché but it needs to be strong) forever! For fiction this will be an extension of your synopsis, tapping into each of these senses. For a non-fiction author you can list the benefits of your information, how it affects the reader and how it will change the reader's life for the better.

You also want to hammer into the prospective reader what is unique about your novel, because we place more value in things that are rare; "That rusty coin you see in the shop window? Might be dirty, but it's the last of its kind. That will be $5000 thanks…." If you can make your novel unique then you are half way to making a sale and gaining a reader. Try to make the offer irresistible. Offer several free short stories based around the world you have created, or offer a free audio interview with an expert in your niche if they buy your non-fiction novel. Offer access to a reader's only discussion board; make your novel an experience!

Also, if you have any reviews or testimonials make sure you place them prominently. The best reviews provide some balance. Remember, reviews such as "this book is the best book ever" are hard to believe at the best of times… Your 'About the Author' section is another chance to show a reader the author credentials from your book proposal. Tell them exactly why you are the best person to write this book, what your experience is and how events in your life have lead you to the point where you have written this book for **them**, your readers. Take a look at the My Story page on the Craven Stories website for inspiration: *http://www.cravenstories.com/my-story/*

What is by far the most important part about your website? That it is **ALWAYS** about the readers. It is in no way, shape or form about you. Ever. You can relate to your readers and share similar experiences but your website is not an avenue for the Me Show. But then, when is it ever about the author?

3.3 Turning Your Once Off Readers Into Return Readers

Many a successful marketer will tell you it takes anywhere from seven to fourteen connections between you and a prospective buyer before they will trust you enough to purchase your product. Granted, no internet marketer I have ever met was giving away their product for under five dollars, or free for that matter. So my instincts say we have the edge here and can probably downgrade that number to around two or three.

The current mindset in our industry is that you sell a copy of your book and let the buyer go, hoping that when you release the next one they will find their way back to you using some sort of sixth sense to discover your new work. This seems like a roundabout way to make an income. This bookish sixth sense is not reliable by any means and seems to depend on how well you wooed your reader when you wined and dined them with your words.

What if there was a way that you could stay in touch and have your readers flocking to your next book when you contact them? "Em, you're living in a fantasy land," you say. "People don't just give their email to random authors – that's crazy talk." The point that many an author seems to be missing is that, as a race, humans like to socialise! It's part of human nature to join communities and be a part of something. You are cheating your readers by not giving them the chance to connect to the community you create and be pulled back into your world when the pressures of life get too much and they've just forgotten how good it feels.

Every reader needs a little nudge. You are wasting your time on all this marketing if you are not giving readers the opportunity to opt into your database so you can communicate with them later. Cutting a reader loose after they have delved into their pocket only once is a bad business plan in any industry. Remember that a reader's attention span lasts all of five minutes. You need to keep cropping up in their lives to be remembered, kind of like a spinster aunt who randomly shows up at birthdays and christenings to prove that she still exists and wouldn't mind a helping hand here or there with the cats.

Readers go through so many books and so many authors that we become forgettable. We merge together in one giant author lump so we look like an unappealing Picasso. What you as an author should be doing is making sure they know you are still here! You're still alive, and you've even trained your cats in the meantime to do some pretty cool shit. Why don't they come and have a look? The important question is this: how do we get them to give their name and email? It's easy really, I'm sure you would know the technique by now.

Stop selling upfront! If you shove your novels at them like a mother shoving greens at a child, you are going to end up with one very messy floor and a really pissed off kid. You need to connect with readers first so they will trust you. And the best way to capture the attention of a reader and their email is to give something for free. Whether it is information, audio interviews, a short article of handy tips, a weekly or monthly newsletter on topic X, or a signed caricature of you turning a dragon into a mouse, it doesn't matter as long as it grabs their attention and they see some value in it. It's important to note that 97% of people will not buy from you the first time, so offering a prospective reader the chance to opt in may also see sales of your book increase.

If you are a non-fiction author, a fortnightly newsletter with handy tips and articles on your niche is a fantastic way to connect. Whether the newsletters are for your self-help book or a book on the world's best nudie runs, the best part is you don't have to create all of the content yourself. A great place to get free articles is *www.ezinearticles.com*, where there are articles on just about every topic (I'm sure nudie runs are in there somewhere) and it is free as long as you keep the About Author section at the bottom of the article. You can disable any hyperlinks attached to the section if you are worried about it being too easy for your readers to access someone else's product rather than yours. One of the great mantras you should constantly repeat to yourself is, 'People do not buy content, they buy convenience'. Though you may be giving away some of your tips for free in a newsletter, the reader will still buy your e-book because all

3.3 Turning Your Once Off Readers Into Return Readers

the information is there in one conveniently ordered location. Another great idea for non-fiction authors is to give away information for free, or part of their e-book for free and then offer their services as a consultant in that field. If you have enough knowledge to write about your niche, you have enough knowledge to advise others.

So give something away for free and then follow up with other emails! "Ok Em, that's all very well and good but I don't have time to email all my thousands of fans that decide to join me." First, let's deflate your ego a bit there; second, this is really a non issue. It is at this point in time that you turn to auto-responders. Auto-responders are your best friend. Anyone who opts in to your 'list' can be entered into an auto-responder. One such auto-responder is Get Response (*http://tiny.cc/EBRgetresponse*). Then all you have to do is write one email which you can customise with your readers' first names (the program uses a little code to input the name for you) and then send it to your entire 'list' (all your readers) with the click of a button. So not only does it allow you to provide mass updates to your readers with very little grunt work, it also allows you to be spam compliant.

Writing a regular email is very easy and you can recycle them for each new subscriber. There are several key factors to remember: don't forget to personalise it with their first name, as this is very easy to do in an auto-responder. By personalising it with their first name the reader feels you are reaching out individually to them. People are forgetful so every once in a while you should put in a brief sentence reminding them of why they signed up. What was the benefit of it? Remember, you are selling yourself and your writing in every email. Your subject line is your headline, a little piece of curiosity to lure a reader into opening your email. Remember only because only half the people you send it to will open it. Your emails need to be short so you should write in a very clear and succinct way. We are all time poor and a reader will thank you for being brief and entertaining. Basically your aim is to make them stop and take notice. Like a well written novel, you need to make sure every sentence has a purpose. One handy tip to remember is if you are offering something for free, say a short story or sample chapter, make sure that you put some sort of character between the letters in 'free' such as 'f.ree' or 'f~ree', otherwise you will be picked up as spam by several email filters.

Do not spam your readers with constant promotions! Your email should be 70 percent content and 30 percent focus on your novels. You are trying to create a connection, don't screw it up by pissing them off.

Having said this, you also need to know that readers are a little like dogs. You have to train them to be committed, as it were, to your community. At the end of each email you want to make sure you have a call to action. Whether it is to get them to click through to an interesting link, check out your latest blog post or participate in a character naming competition. You need to get them to take some sort of action at the end of your email. It is recommended you put this link in at least two or three times. Get them use to taking that action, and connect your links with things that amuse and entertain them. Then when it really matters, like when you are launching another book, they will willingly take a look at the book and purchase it. It is only by converting one time readers into return readers that you will start to see real e-book success. But nurture your connection carefully, because the readers can bite…

***

I dive in depth into this topic in a short mini-course on my website. I remember the days when I had to DIY it because I didn't have the reputation I have now. So this course is for you guys, the boot-strappers! Those who know that knowledge is power, and Google is a time suck. The course is designed to empower you while at the same *not* break your bank balance and contains step-by-step videos that lead you through everything you need to know to start your list. You can find out how to create your own email list here: http://www.cravenstories.com/books-and-more/courses/

3.4 Reviews: How Credible Is Your E-Book?

So you have your first five star review, or five reviews, or ten.... It depends on the number of family members you have. Unfortunately, unless your family member is a well-known book critic, chances are people WILL doubt that all your twenty plus five star reviews are genuine and not just your ego going for a joy ride under a fake name.

You need to remember that the best review is a balanced one. If you are deleting all your bad reviews and leaving only squeaky clean ones, a little switch is going to flip inside the prospective reader's head flashing a red warning light. The readers aren't stupid; they know when you've been rigging the results. It's important to note that a one star review is not necessarily always negative. Indie author Brian Pratt, who made at one point almost $20,000/month with his e-book sales, noted how one of his bad reviews led to generating one very happy fan. The review in question was along the lines of, "This is horrible, it reads like a D&D (Dungeon and Dragons) videogame script." Brian received an email from a fan – a large player of videogames – who cited this review as the reason why he tried and then bought all seven books in that series. 'Bad' reviews should never be discounted because one reader's dislike is another reader's passion.

If your friends or family members do write reviews of your book make sure they include both the good points and the possible problems that people might have with the novel. They should also pick out quotes from the novel to prove their point if possible. Basically they have to say exactly why they enjoyed the book. What did it accomplish? Did the characters stand out? How did they stand out? Was it a thrilling spin on an age old theme? Then have them note what points might not appeal to some people. For example, "It's great for fantasy and sci-fi lovers but if you aren't too keen on romance this might not be the book for you." One person may have trouble believing a dog can talk while another reader might say, "Bring on the dog humour."

Reviews are supremely important to the credibility of your work. A good spread of reviews will let a reader know what they are in for. It is a recommendation from one reader to another, rather than a sales pitch from an author to a reader. However, not all reviews are made equal. A review from an influential and respected reviewer is needed to really make an impact. One of the problems for self-published indie e-book authors is that few reviewers accept a) e-books and b) self-published books. This is mainly due to the reputation of self-publishing in the industry. A few people were a little overzealous with their substandard novels and buggered it up for the rest of us. That's life. It may be possible to set a new standard though. I am a large believer in the theory that unless you give it a go you won't know. So I would suggest to you, if you have a good quality book (that has preferably been through an editor) that has been well-received in the e-book stores, then think about sending an enquiry letter (which I will discuss) to these reviewers. Let's see if we can help shift them to a more open minded view of e-published works.

Reviewers generally read genre specific books and so, in a sense, finding reviewers is kind of like finding a publisher: you have to be precise. So to find a reviewer in your genre or niche all you have to do is type 'book reviewer' + your niche into Google. While you're searching, find out what they have reviewed. You don't want to send a paranormal romance to someone who reviews paranormal thrillers. Are the books they review ones that you know? Do they relate to your type of genre? How many people follow their recommendations? Do they review for publications? There are many people who will offer to review your book, but only a few will have the public presence that you are looking for. Or you can look at books that are similar to yours on Amazon and see if any of Amazon's top 500 reviewers have reviewed it and if there is contact information for them.

Then you need to check out the reviewer's submission guidelines. Some reviewers ask for you to send an enquiry first, others will tell you just to send the damn thing. Each reviewer will have slightly different rules on what they will and won't accept. Very specialised non-fiction filled with jargon may need to be reviewed by peers with a good reputation rather than book reviewers due to its technical

3.4 Reviews: How Credible Is Your E-Book?

nature. Reviewers will not read unpublished manuscripts. Remember, as a self-published author your book has to be of exceptional quality for major reviewers to consider reading your work. Reviewers do not mince words; they will not recommend your book to anybody if it does not read well. Generally they prefer that your book is currently available to the public (although some, contrarily, want your book several months before you release it). Some e-book reviewers will read books published by indie authors (including small press authors) as well as traditionally published authors who are re-releasing backlist titles on their own, so that's something to keep in mind for authors who are out of print.

However, if you want to appear professional and stand out from the hundreds of books that reviewers get every week, there are several things you must do. You should ALWAYS attach an enquiry or cover letter to the body of your email to the reviewer. This needs to be short and to the point. Give them your first and second pitches; tell them your author credentials and why these credentials qualify you to write that book. Mention at least one of the books that they have reviewed and why their review led you to think they were the right person to send your book to. Finally you have to give them the unique selling point of your book. Why is it different, why is it life changing? This will come from the synopsis you have written and the content you have carefully placed on your website.

For example, indie author D.N. Charles is giving his book away for free to any woman in the Australian sex industry in the hopes of finding the woman who inspired his novel. This creates interest and curiosity. Remember, you are pitching your novel to them. Reviewers receive huge numbers of books and are under no compulsion to read them, so it is your job to convince them to read yours as soon as possible. If you want to approach a reviewer who does not accept e-books or self-published works then send them the query letter only, with the best headline you can manage in the subject line of the email.

Send them your most up-to-date, corrected and perfectly formatted copy! You do not go to a party half-dressed, so why would you send your novel to a reviewer half-polished? Similarly your cover had better be professional. If your cover is amateur you are giving the impression that your work is amateur. Again, you wouldn't go to a party half-dressed, and you certainly wouldn't go to a party in a garbage bag. There are no do overs. The copy you send first is the one they are going to read. So you'd better make sure that you send them the right one.

Send it to them in the format they request whether it be Kindle format, PDF, whatever. If possible, zip the file so that it does not clog up their inbox with its size. You want to make sure it is as easy as possible for them to access your novel. Don't make them download it from another website! They are giving credibility to your work, so expecting them to click on this link or that link and then download it using this or that coupon is a) not a good way to get into their good graces and b) disrespectful and treats them like your maid. The best strategy is to add the e-book as an attachment to your dazzling query email.

You can, of course, go the paid route to get reviews for your books which, depending on the service can yield a large number of reviewers for less research time. However it can cost a pretty penny, but many see it as worth it. Services like Hidden Gems (*https://www.hiddengemsbooks.com/arc-program/*) send your ARCs (advanced reader copies) out to their mailing lists of reviewers and you are charged $3 per reviewer who picks up a copy. Not all of these reviewers will review, but the website itself boasts an 80% review rate from those who do select your book to read. The catch for the service being that you sometimes have to wait up to 6 months for a spot in their calendar.

Another, more expensive possibility is websites like Netgalley (*https://www.netgalley.com/*) which can get you in front of thousand of reviewers, many of whom have large followings and provide incredibly detailed reviews and feedback as noted by indie author Luke Gracias (*https://fiveplustwoblog.wordpress.com/2017/02/03/netgalley-for-a-self-published-debut-author/*). Packages to list your books there can be upwards of $600 USD. Be warned though, Netgalley might be one of the largest concentration of book reviewers, but they are also reputed as being the most *honest* pool of reviewers so you need to be

3.4 Reviews: How Credible Is Your E-Book?

confident your book is the best it can be before sending it into the valley of the truth tellers!

So, it's time to ditch that garbage bag for a cocktail dress (or suit) and start submitting to your chosen reviewers. Perhaps save the champagne for after the review…

3.5 Handling Feedback Is Like Juggling Tables

Have you ever heard stories of what happens when an author blows up? I mean majorly blows up, as in questioning the heritage, parentage, sexual and intellectual capacity of a reviewer who gave a so-so review of the book. They start imagining gremlins eagerly awaiting a chance to dash their hopes to pieces and tear apart their fragile victories.

Perhaps you should check out *http://booksandpals.blogspot.com/ 2011/03/greek-seaman-jacqueline-howett.html*. It's nothing if not entertaining, particularly with the initial use of words such as 'boo boo' and ending with the sentiment 'f***off'. But one thing is for certain: this author destroyed any chance she had of ever getting traditionally published and more likely than not lost a lot of readers over her very passionate and, dare I say, obnoxious rant. A rant that went viral on the internet...

The fact of the matter is this folks: you are never going to get a perfect strike rate when it comes to reviews of your e-book or product. There is always one person here and there who is convinced that you have taken them for a fool and are about to run off with their life savings. One book will never be loved by everybody and that's what makes this industry so exciting; there are so many diverse tastes and everyone is looking for something slightly different. I am just waiting for the time when someone comes back to me and says my advice is about as accurate as a blind man playing darts. Then that day, I'm going to have to deal with it.

How you take bad reviews and complaints in the comfort of your own home – raging, crying and smashing plates like a Greek dancer – should not be how you address them publicly. They need to be dealt with in a professional manner in order to increase your credibility with future readers. Responding to negative feedback is a very powerful way to market your book. By asking the person why they did not like your e-book, thanking them for the feedback and then commenting that you will use it to improve your novel sets you apart from other writers.

In this way readers can see that you can actually take negative feedback and that you might even use that negative feedback to become better. You show that you are not perfect and you can be better, and that you are willing to work to improve.

The trust of a reader will come more to the person who has negative feedback as well as positive, and to the writer who handles that feedback well in public. Even if they are slowly destroying their houses and smashing their hardwood tables to splinters in private.

3.6 You May Have The World's Greatest Book But, If No One Reads It...

... You have the world's greatest Paperweight!

What do you have more of, time or money? I suppose another more pertinent question is, how much of a cheapskate are you? Because this is going to impact how you try to get traffic to your e-book pages. Unfortunately traffic (aka the number of people visiting your e-book download pages) is not organic, does not grow on trees, does not magically manifest from a spaceship in the sky like Captain Kirk and it most certainly does not occur in any significant quantity from you sleeping on the job. Traffic has to be worked for, like everything else.

There are three types of traffic: paid, borrowed and free. While I am going to spend a little more time on the 'free' option (yes, I did just see how much your eyes lit up. See how well the word free works?) in a later section, I will briefly explain the others here.

Paid

So you have the mullah but not the time. Understandable, we are all time poor in this day and age with too many distractions coming at us from all angles. You can do paid advertisement in many ways, through Amazon ads (*https://advertising.amazon.com/kdp-authors*), Facebook ads (*https://www.facebook.com/business/ads*), Google Adwords (*www.adwords.google.com*), the list goes on.

Basically, you advertise on these sites and every time someone clicks on your ad you pay anywhere from cents to your first born – depending on how far up the list of ads you wish to appear – per click. The more money you pay per click the higher your ad appears. When a searcher types the keywords you have selected for the ad into a search engine, your ad appears. This is where researching your keywords and knowing your competition comes in. If you can advertise for keywords that have lots of people looking for them, but not many people selling with those keywords, then you will save a ton of money. And we are all, really, penny pinchers at heart.

There are a couple of effective tips you can use to make the most of your money and draw in a greater number of readers. All three of the aforementioned platforms allow you to rotate through up to over a dozen different ads (different headlines) at no extra cost. This is a fantastic option, because the platforms will tell you which ad gets the most people clicking on it and you can eliminate any ad which isn't drawing your readers in. I discussed in a previous chapter how to set up headlines for your website and novel to entice readers. You need to think of your ad as one big headline. You don't need to sell your product in the ad; you just have to get the reader curious enough to click. You need to keep your ads short and sharp. Rambling is not effective.

If you are a non- fiction author, think about the dozen main benefits of your e-book and turn those into ads. Make a clear promise to your reader: what is in it for them? And you can, as always, make it irresistible if you offer something for free, whether it is a sample or your whole book. Also try using your keywords in the ad. If your reader has typed in 'Ninja Secrets' they are more likely to click if they see those words in the first line of the ad. And always test your headline. I have heard stories of ads which barely had any clicks, such as "Put music in your home," being changed just slightly to, "Puts music in your house" and making millions. You test to find your Superbook combination and you test to find your super-ad combination also.

In all cases, you need to decide how long you want to show your ad for. If you advertise using one set of optimised ads for three months constantly, the users will get use to the ad being there and they will go blind to it. Once you go chameleon and blend into the background, it doesn't matter what your ad contains. It could include a photo of naked women doing the CanCan on the back of a pink elephant and it wouldn't make a difference. So it's important to do it in a pattern where you advertise for several weeks, then remove the ad from the site for several weeks and repeat. If you chose to place an ad on websites or blogs of individual bloggers or reviewers, always independently check to see how many visitors per day a website is getting. You want at least

4.6 You May Have The World's Greatest Book But...

500-1000 people minimum per day frequenting the site where your ad is placed.

Make sure you also track how many people are clicking on your ad from each site. Keep the ones that give you a high number of clicks and get rid of the duds that give you minimal traffic.

Borrow

Also known as joint ventures. Here is where you do your shady deal with the crazy man in the alley. Well, not quite. What you need is someone in your field – if you know the person it makes it easier – who has a large database of people they can contact. Then you do a deal with them to split the profits of your sales if they email their database about your e-book. This works better for non-fiction authors who know of other experts and sub-genres within their niche. If you have a buddy who is a lecturer at a university, capitalise on it!

On the other side of the coin you can look at this as writers helping writers. If you have a fellow indie author in the same genre whose novel you have read and would recommend, you can collaborate and borrow each other's readers/traffic. Put their synopsis and a mini version of their cover in the back of your novel and have them do the same for you. Then every reader who finishes your novel, full to the brim with praise for your prose, sees the ad to the book. Because they liked your book so much and it is in the same genre as yours, they take your recommendation to purchase the other novel. It is a win-win situation for both authors.

If you have an Amazon account you can also use an affiliate link to your buddy author's book you put in your back matter, that way you get a piece of the sale if the reader comes through the link in your book. Then not only are the authors helping each other out, but they are also getting a little commission (or percentage of the price) for recommending the other author's book.

Free

Three for free! Traffic methods, that is. The first two I shall mention briefly as I will be covering them in more detail in later sections. The first is social media, a topic you have all heard so much about. By connecting with people in your niche through such sites as Twitter and Facebook you can generate large amounts of traffic through promotional posts. It is always important to balance the promotion with sincere participation within your community, with promotion only taking up 30-40% of what you write. The more you engage with your followers, sending them to funny links and informing them on your topic, the more impact your promotions will have. The second is writing articles on your topic. This is a great way to get promotion without having to do the leg work.

The third and final way to get free promotion and traffic is by creating a press release. Have you ever wondered how the footage of a budgerigar riding a surf board got on the evening news? Or how every newspaper seemed to be present when Richard Branson was making one of his 'surprise' stunts? Most articles in a newspaper are there because someone sent a press release to an editor. This is one of the best kinds of marketing, because people trust media outlets more than your advertising. If you can make your e-book newsworthy then you can get free publicity. However, do not make your business the focus! That will get you discarded faster than a publisher rejection.

There are three ways you can make yourself or your e-book newsworthy. A) Do something good for others (charity); give in the right spirit and the publicity will follow. B) Make announcements that affect a lot of people or will intrigue a lot of people. In a lot of non-fiction cases a unique product or e-book release will particularly interest publications of your niche market. C) Have a strong and clear view on a news topic. If you have a strong opinion and can back it up, say it in a press release. Editors love opposing sides of a story so keep a close watch on internet sites related to your niche for topical subjects.

A press release is similar to a synopsis: it must intrigue and excite, tell the story quickly and concisely and provide contact information if an editor would like more info. For example, internet marketer Brett McFall had a four day summit on internet marketing. Though it is a very informative event, it was not enough to get an editor's attention. So he made an announcement before the event about a previous attendee who would be speaking and how this attendee had made $96,000 in five months after going to the event the previous year. Indie author D.N.Charles offered his e-book for free to any sex industry worker in Australia in the hopes of finding the one person who inspired the novel. This story made up the basis for his press release. Websites such as PRWeb (*www.prweb.com*) or if you're in Australia Medianet (*www.medianet.com.au*) are several places where you can send your press release to media outlets. But the people most likely to cover you, are your local media, so don't forget to track down the journalists for your local TV station or rag to send a release to!

Pick at least one of these methods of traffic generation to start with. Do not leave your book out in the cold unsupported!!! Because the fact of the matter is this: you can have the world's greatest book, but if no one reads it, it becomes the world's greatest digital paperweight.

3.7 The Road To Free Traffic Is Paved With Articles

As writers, writing should be our passion, from recounting five inch goblins storming a giant's castle, creating detailed to-do lists, to describing in exact and colourful language why your friend is such an arse in his latest Facebook photo. So I know you will just jump at the chance to write another several pieces of mastery in article form. No, don't look at me like I'm a slave master with a whip. If you want free traffic, you've got to put in the effort; anyone who tells you they're where they are today for free must be fresh off the funny farm.

The main goal of the internet is to provide information and entertainment; this is what people search for. So to get free traffic you need to give away some of your knowledge or provide a little entertainment. The more you teach others what you know the more information they will want from you. What you need to do is give them good quality information or a well-written and entertaining analysis so that the reader will visit your site. Don't be lazy and just write one; flog this method for all it's worth because very few writers know these avenues exist.

Once you have written several articles complete with an About The Author section – including your web address – you publish them on sites (for free) such as *www.ezinearticles.com,* or websites run by other bloggers who are prominent in your field (or on their way). These websites allow people who want content for their newsletters or blogs (generally for their own gain) to use your article, but only if they use the entirety of the article, which includes you author information, e-book name and website address. These articles can potentially get sent out to thousands of people.

You want each article to be at least 300-500 words. For fiction authors you can do reviews on other books in your genre or write articles on events, news or trends in your genre or even a How To Write guide for genre xyz or audience xyz. For a non-fiction author, commenting on themes in your book or giving away tips and information are the best ways to go. The article needs to be similar to how you have written the content for your website and it needs to be very clear on what the benefit is for the reader; make sure that it is based on their interest. Non-fiction articles need practical information that can be used straight away by the reader. It should NOT be a sales pitch for your e-book. No one will use your article if it is, only the insane give free advertising for nothing. It needs to stand on its own as a helpful bit of information or entertainment; if you want people to use it for their newsletters, it has to deliver.

If your information or tips are from your e-book this can also be a great way to advertise without obvious promotion. In the About The Author section you can note that the article was taken from the new e-book XYZ and for more information they should visit your website (don't forget to put your website address there!!!). Your about author section should be concise, giving your name, genre and second pitch for your e-book and an invitation to check out your website for more information, or more enticingly, to check out your website for free samples, free short stories, audios or whatever you would like to give away.

All you need is for several newsletter 'publishers' to include your article in their newsletters for people to start visiting your site. If you find you can't write several articles in one sitting, break it up into manageable chunks and write one a week and you will get all the traffic you need without paying a cent.

One extra avenue for your articles is to submit them to Digg, where people can rate (or 'digg') articles based on their content and entertainment value. A lot of bloggers and social media addicts use Digg articles as content for their news that day, and a well-written article can see you climb the rankings of Google very quickly through exposure on Digg. Articles that rank best on Digg are ones which are funny, educational and entertaining in an extraordinary way or which turn a topic into something controversial. Basically you need to write about something people want to read (generally topical) and link it back to your product at the end. You can also climb the rankings in Digg with a funny YouTube video (something I will touch on in a later section).

3.7 The Road To Free Traffic Is Paved With Articles

Don't forget to use these articles as content for your blog as well. As mentioned before, blogs climb up the rankings of Google much more quickly than a website because Google likes fresh content. Posting on a blog at least once to twice a week is a great way to stay high in the Google rankings. And at the end of each article you can put a link to your author website, if you have one, or your e-book.

It's also good to note that one way of finding ideas for your next article/blog post is to sign up for Google Alerts. There you can specify keywords or search terms in your genre and every time there is any news related to those search terms Google will inform you. As such you can write articles about the most up to date news in your genre/niche and link it to your e-book at the end of the article.

So, it's time to check out of the funny farm and get to work!

3.8 Consider How The Humble Short Story Can Increase Your Fan Base

We are living in a world of entertainment bites. When we are bored on the commute to work, waiting at the doctor's surgery or stuck in rush hour we want something that passes the time and resolves itself quickly. We are addicted to fast pace storytelling and, like a drug, once addicted we keep coming back for more in larger quantities. If you can tap into this addiction and become an entertainment bite 'pimp', you can quickly and easily build up a fan club for your work. Readers get a sense of your writing style, your characterisation or your expertise from these little snippets of entertainment. Put in links to your full length e-book at the end and suddenly you are getting more downloads and readers. When I talk about entertainment bites I am of course talking about short stories.

If you have never thought of short stories as a marketing and connective tool, think again. Publishing giant HarperCollins has already jumped in there with both feet, branching out into short stories as a way to tap into a large part of the population who does not have the time or the interest to read a full length novel. Several short stories by authors such as Tobsha Learner are already available through Amazon and Kobo, retailing for a cheap but profitable $1.47. Self-published authors need to take part in this emerging trend where current technologies, such as iPhones and smartphones, play a significant part in our readers' lives. In this arena we have the advantage; we do not have the overhead costs that HarperCollins has to cover. We can sell our short stories for anywhere from 20cents to 50cents or give them away for free to draw readers into our world.

Short stories can be anywhere from 500-2500 words in length. They generally leave an unanswered question or a curiosity at the end. Within the piece you must be able to swiftly sketch the situation and come to the point. There are fewer complexities than in a novel and the story focuses on one situation or scene, or for non-fiction it focuses on one aspect of a theory or opinion. By linking your short stories to the world of your e-book and exploring extra bits of information that were not included in your novel, you can use these pieces as a promotion of your e-book by creating interest and anticipation around your larger work. It is also a great opportunity to explore characters that only feature in a minor way in your e- book. It gives the reader a little thrill of discovery, another piece of the puzzle. Short stories need to be self-contained and, though they may be part of the bigger world created in your novel, they need to satisfy the reader as a standalone piece as well as a part of the whole.

For non-fiction authors: consider putting out short reports or case studies. These should either be free or very cheap (as in 20cents). You want to engage your reader; you don't want them to feel as if you are ripping them off. Don't try to profit from these minor works because they are just a way to further connect and showcase your writing.

If you only have one e-book and not a series, free short stories can be just as effective in boosting your sales as giving away the first book in a series for free. If you can entertain, move or engage your reader with a shorter, cheaper work, you will connect with your readership on a level that many writers do no manage. You reach a larger portion of the population and can hopefully convert them to join those of us who read longer works for pleasure.

3.9 The Power Of Recommendation: Contributing To Your Community

If you don't have a notebook, buy one. It doesn't matter what it looks like: short, fat, large, thin, pretty, recycled, a couple of old envelopes stapled together, whatever. No, I will not be getting you to write lines. Today we are delving into the world of blogs, forums, discussion boards and groups. Chances are that for every niche out there and for every genre you can think of there is at least one community built around a love for that topic. People who like to wear funny hats, check. People who feel going commando is a better option, check. People who enjoy dancing like Mr. Bean, check. People who watch Tom Cruise movies obsessively, check. Chances are you have found several while researching your market. One of the best ways to directly communicate with your audience is to participate in these communities. It also helps boost the popularity of your website with Google. Aha! Now I have your attention.

Studies have shown that 78% of people trust the recommendation of their friends or trusted community members over advertising. By becoming a part of a trusted community (or several, hence your need for a notebook to keep track of all your user IDs and passwords) your words hold sway. If you make a significant contribution to the community, such as a relevant comment, post or question and then finish your contribution (sign off) with a link to your website or e-book, participants of those communities will be more inclined to click on your link. This is another case where you must stop selling your novel upfront. Get people to opt in on your website (through your sign off link) so you can convert them into readers at a later date, or send them unobtrusively to your book by making the link easily accessible. If you are not a contributing member for the community and are merely promoting your e-book you will, more likely than not, be banned from the site. You will be seen as SPAM, and not the meat product of highly dubious origins, but a person who fills up a person's email or website with 'junk'.

The best way to find groups in your niche is just to Google your genre and the words 'forum' or 'discussion board' or 'group'. There are Google groups (*http://groups.google.com*), Yahoo groups (*http://groups.yahoo.com*), Kindle boards (*www.kindleboards.com*), discussion forums on Facebook, Goodreads, blogs and websites. Look up authors or experts in your genre, and see if they have fan sites created by themselves, their publisher or their fans. Generally there are discussion boards and forums present on these sites that you can contribute to. Target the online communities where your readers connect. As I said, make sure you contribute to the community then sign off with your name and below that the link to your website. Your comment can be anything from recommending books, discussing the traits of a fictional character, the merits of a theory or just agreeing and adding to the response of others, any of these are valued. If the comment board will not allow you to add the full link to your e-book or website, just put in a text version of it – *Author of E-book Revolution*, for example. But your first try should always be with a clickable link. People are like pandas, they are lazy, so you want to make it as easy for them as possible.

This is also a great way for authors to tune into the trends in their genre. You can find out what people want to read, what questions they are asking and what they need. Non-fiction authors can ask questions about the problems and frustrations of their potential readers and hence find out how much time they should devote to different sections of their novel. A non-fiction author can even show people how to turn their hobby into a business. Fiction authors can find out what sorts of traits their readers want in their heroes, who their favourite bad guys are and why they prefer this mystical dragon over that one. The opportunities are endless.

The added plus of posting on forums and communities is this can get you higher up the Google ladder. By placing the link to your e-book or website at the end of each post or comment you are creating what is called a back link to your site. The more back links your site has across the World Wide Web, the higher up in Google your site

4.9 The Power Of Recommendation

appears. Google rankings also depend on other factors, but back links are a major one. That is why it is so important to place the link to your e-book in every online activity you do. But for this to work well, don't provide links to multiple sites or e-book stores that stock your book, instead send everyone to one centralised place (this is where websites and blogs come in handy) or e-book store. It takes a LOT of back links to make a dent in the Google ranks, so every post counts!

However, don't just concentrate on the technical side of things – have a little fun! Make some jokes, entertain, educate, and engage. If we don't find enjoyment from our work then what's the point? Certainly not to pretend you love base jumping when the thought of climbing two stairs to the front door makes you faint.

3.10 You Want Raving Fans, Not Rabid Ones!

Creating an online community is similar to creating a hippie commune. You gather a bunch of like-minded people, give them a task to keep them occupied, ration the weed, and appoint yourself mayor to 'supervise'. Hmmm, no that's not quite right. Let's try again. Creating a community is like being the dictator of a faceless reader army. You train them to eat, drink, read and crap exactly as they're told and then whip them into such frenzy they rip the other authors' poorly entertained armies apart! Nope, that's not it either. Sigh.

Finding the right way to truly connect with a community is hard, yet there are little touches you can add to engage your readers and give them a sense of place and a feeling they are part of something greater than themselves. A community doesn't evolve by handing your novel over and taking a lazy back seat, nor does it evolve by you donning a Hitler moustache and trying to dictate how your work should influence a reader's thinking. Though contradictory in some regards, reading provides a social opportunity for discussion of literature. Smart writers help cultivate that discussion.

Fantastic sites such as Goodreads (*www.goodreads.com*) exist where people can discuss the books they are reading with their friends. It's kind of like a book club mashed with Facebook with a little bit of e-book store added into the mix. It has so much potential for creating a community. Readers can add notes to the margins of books, highlight text, and discuss their favourite aspects of the e-book with other readers in the community. Similar to Facebook, you can start your own groups, whether they are for book clubs, fans of a particular genre or around the books of a particular author... such as yourself perhaps? Are you starting to see the potential? This is the future of the book club. You can search for people who read in the niche markets that you are aiming for. From there you can recommend your book to them, invite them to join your groups or 'follow' them. You can answer the questions readers add in the margins of the book and fans can discuss your work with other fans. You can construct reader group questions around your novel. You can point your readers to these Copia groups so they can become part of your wider community without you having to deal with the reality of setting up your own forums or discussion boards. Guiding readers gently to these experiences gives you another link to their life, and the more links you have the more likely they are to keep coming back to you.

Some authors have already grasped this concept of community and merged it with fan fiction. The basic premise is that the author starts a large discussion board with various character scenarios based within their world, and lets their readers' imaginations take its course. As part of the conditions for readers to participate, readers allow the author to use elements of their fan fiction in the author's actual novels (for no monetary reimbursement) however the reader's name will be acknowledged in the book dedication. So not only does this create a community around your work allowing the readers to step into your story and create, but it also adds an element of excitement for the reader. Not only is their idea deemed worthy by the author but they are also elevated above other readers for their effort AND they have their name in print.

Cory Doctorow is one fantastic example of how to make writing a more interactive experience. If a reader finds a typo in one of his books, on the next print run he fixes the typo and puts a footnote on that page in the next edition thanking the reader for spotting it. Not only does this acknowledge the reader and give them the excitement of appearing in the next edition of their favoured author's novel, but it improves the quality of the e-book and makes each print run unique. An author should take every opportunity to make their book as unique as possible.

Author Jim Brown makes the offer that if you join his mailing list not only will you get his latest novel for free (here you could put in a short story, report or tip sheet depending on how much you want to draw the reader in) but, if you sign up before his next novel is published, he will use your name for one of his characters. How excited and juiced up is that reader going to be to join the mailing list? Very!

Not only is it your job as an author to engage, it is also to facilitate

3.10 You Want Raving Fans, Not Rabid Ones!

an environment where your book takes on a life of its own. It's time to loosen that death grip you have your reader in, seriously, it makes you look desperate. The more creative you are with your experience, the greater the buzz that will surround your work. Don't go overboard though – you want raving fans, not rabid ones.

3.11 Facebook and Twitter: The Gateway To Thousands Of Readers

It's a surprise that we haven't all turned into light hating golems after spending so much time on the internet. There are 1.9 billion of us online and over half of us are perched on the edge of our seats throwing virtual cows at friends and poking random strangers. Social media is a phenomenon that has gripped an entire planet across an amazing breadth of ages and interests. Social media is how you connect with people, who think the same way you do, across the globe within milliseconds. It is the most powerful tool to connect with people on the other side of the world who would normally never get the chance to read what you have written.

Many people um and ah about whether or not they should use Twitter (*http://twitter.com*), or use Facebook (*www.facebook.com*), or Instagram or Pintrest… the list goes on. But in the time that you take to research and have an in depth conversation with other writers, you could have just set up a Twitter or Facebook account and road tested it. Social media is not something to be afraid of; we are not connecting with Martians in space. It is also going to be how we communicate in the future. If e-books are a digitisation of reading, then social media is a digitisation of communication. They go hand in hand and if you want to be successful online you are going to have to brave one or the other. You can sit on the fence all you want, but while you do the rest of the world is powering ahead. Let's go through some great ways to get started on Twitter and Facebook specifically, and how you can target the right audience.

Twitter

Setting Twitter up is as easy as typing in your name, email, and creating a user name and password. You can set up your profile with a little bio and picture and also add your website URL into the mix. Apologies if this first part is a little simple, but I want to make sure I provide the info for those who have really been resisting looking into it! On Twitter you send out what is known as a Tweet, which is a text based post or comment of up to 240 characters in length (links to websites are included in the 240 characters). The principle is similar to a blog except here you need to be more precise and entertaining; building up curiosity is the key. As I noted before, it is important that only 30%-40% of your tweets are for promotion and the rest need to be building up credibility, connecting with followers, sending people to interesting sites and being an entertainer. However, with this sort of entertaining you don't have to contend with stage fright!

When you are writing a tweet you need to be in the shoes of your client. It may be a bit uncomfortable if your readers are children (such tiny feet...) but bear with me. You need to entertain them, either by providing value and content or quotations and jokes relevant to your genre and niche. You can provide content in the form of articles you have found on *www.ezinearticles.com* or on *Digg.com*, or just by typing into Google 'Tweets on <insert your niche here>'. Another great website that has really topical news and human interest articles is *www.alltop.com*. When you do your 30% promotion don't be blatantly obvious about what you are doing. You need to create interest and curiosity. For example, if your genre was in public speaking you could use a tweet like this: "People fear public speaking more than death. At a funeral people would rather be in the coffin than say the eulogy <insert URL>." It creates intrigue without appearing to promote too heavily.

Tweets are also better if you do it from a third party point of view so, rather than saying "I found" or "I wrote", talk about how your 'friend' found this great book. If you see someone Tweet about your e-book, ReTweet the Tweet using the ReTweet option at the bottom of that person's Tweet. There are two reasons for doing this: a) people love to see their Tweets ReTweeted, and 2) the more people who give this recommendation for your e-book the more credible it appears. You can also band together with other authors and Tweet about each other's books with Tweets such as: "This guy seems to have the best free book I've seen so far <insert URL>." Just make sure you're convinced of the quality of the book first! You don't want to be hawking an e-book that looks like it's been written by a six year old.

While you can have great fun on Twitter, chatting between friends, being on top of the news and trends or just swapping ideas with people, you must make sure that you target who you are following. There is no point connecting with a brass band lover in New Orleans, or an exotic dancer in Russia, if you are trying to sell an e-book on scuba diving and they are in no way interested. Here is where the leg work kicks in. While social media is fantastic for connecting, you cannot wait for your readers to discover you; you need to seek out your readers and interest groups and show an interest in them first.

You want to follow people who will, a) follow you back (so that they see your Tweets and hence your promotions), and b) who have an interest your e-book's genre. The best way to do this is to Google experts or popular authors in your niche to see if they have a Twitter account. The people who are following them are most likely fans of the expert/author or of their work. If you then follow the people following this expert or author the majority of the time they will follow you back. Another avenue is to go to *http://search.twitter.com* and type in the keywords that relate to your target audience and niche (as harped upon countless times in previous sections!). The search will then come up with hundreds of users who are typing those words into their tweets. If the people who are following you are not interested in your genre, they are about as much use as a tutu without a ballerina.

Facebook

The set up process for Facebook is exactly the same, the only difference being that you have to provide both an email and a phone number to verify you are a real person and not a faceless automaton. To keep your personal Facebook Account separate from your business one, you set up what is called a Facebook Page. Facebook allows you to do a whole host of extra things in comparison to Twitter such as add interests (or your keywords) to your page, your philosophy and your bio. These are not restricted to 140 characters like they would be in Twitter. Be careful with your personal information; I wouldn't recommend sharing things such as your address or phone number. However, you should make sure that you have all your websites listed and an email, separate from your personal one, where people can contact you. Being aloof will not get you the following and sales you desire; you must connect with people and give them every opportunity to connect with you. Your main way to connect on Facebook is to post a 'status update', which is sent to all of the people you have befriended on Facebook, or to write on a person's, group's or page's 'wall' (i.e. their profile).

As with Twitter, you use the search function at the top of the page to search for groups or 'pages' that are based around your keywords and the interests of your users. There may be several groups and pages with the same name or phrase but each one will vary in the number of people who like a page or have joined a group. Join the groups or pages with the highest number of users (at least 300 people) so you can reach more people, then make your presence known by posting on the 'wall' of these pages. Think of them as just another forum board. The people who participate in these pages will have an interest in your niche and hopefully your e-book. The more you participate on these pages, the better known you become. Start inviting people to be your friends or like your page if they interact with you on posts (and only if it feels natural, don't force it!).

Your Facebook page (it should always be set up around yourself as the author, rather than a separate one for each book or series you write), allows people to connect with you about your work and creates a community. It is also where you put direct promotions for your e-book. A page is so you can turn that trust into a willingness to buy. You should do at least one status update a day that isn't promotion. You can then add a Facebook button to your blog or website so that people can connect with you easily via social media.

It is amazing how social media absorbs people's lives, and if you can find a way to contact those who care about what you write, on a platform with almost 200 million users, you will be laughing so hard relatives may have to call for the straight jacket.

3.12 Go Viral With YouTube

The internet is like one large incestuous family. The amount of apparently separate companies owned by Google is just mammoth. There is a fair amount of monopolising going on; however, this is something savvy authors can take advantage of. Not only is YouTube (*www.youtube.com*) the world's second largest search engine, and it has the second largest user base of any social media platform after Facebook, it is owned by Google as well.

YouTube videos rank very highly on Google (favouritism much?) and because Google owns YouTube it applies a little bit of parental bias to the site. So if you can pack the title of your video, the title of your upload file (yes Google even checks the words there), the description and the tags full of all the relevant keywords relating to your e-book and target audience, chances are your brief little video will get higher rankings in Google than any website you create. Place the web address for your e-book download in the first line of the description and you have another fantastic link in the marketing chain for Google to find. This is the reason why many authors these days are creating book trailers for their novels.

In a previous section, I spoke about the power of having a YouTube video or audio of yourself welcoming people to your site and giving a quick pitch for your e-book. YouTube can be used for so much more than this. You can produce your own video trailers for your novel, or have your fans create videos of them acting out scenes from your novel. You can make your book an interactive experience by recording yourself reading snippets from your book. Then place a hyperlink (as in the web address to the video) in your e-book text so readers can click on the link and actually hear you presenting that excerpt to them.

YouTube is also the best chance that you will ever get for your work to go viral. By 'viral' I mean that someone watches your video and is so hugely entertained that they email the link to all their friends and post it on Facebook. Those friends then view the video and tell their friends and the effect snowballs. Cast your mind to all the entertaining ads we see on TV that don't have anything to do with the product being sold but stick in our heads like glue. Or all those *Jackass* style videos that fill the *Funniest Home Video Show*. These are the sorts of videos that people pass on to their friends and you can make these types of videos your own. Videos of kids having tantrums while you zoom up on the mother's harassed face can be turned into golden moments. All that's needed is a simple voiceover near the end of the video proclaiming, "If only I had blown Harry off for a night of reading <insert novel title here>." And you have viral comedy attached to your e-book.

While it's best to tailor these videos to the interests of your niche, if you entertain enough people – whether it is with a crude video of a man crushing his genitals as he leaps over a pole or a woman who slaps people every time they swear – your e-book success could skyrocket more quickly than you ever imagined.

***

Are You Missing Out On A Publicity Goldmine?

The power of a simple reading should never be underestimated. Hearing the dulcet voice of another verbally caressing your prose seems to add another dimension somehow. As much as we try to deny it, some little part of our inner child is still fascinated with the daring escapades of talking shoes or pink elephants rendered vividly to life by a story teller's fairly dodgy accents and high-pitched voices. The simple act of giving voice to something that is normally just in our head – particularly comedy which always reaches its true potential when performed – deepens our attachment as voices and sounds are added to the pictures of our imagination. And creating a clever video round your e-book could see it sky rocket to the most talked about digital file (or stapled pieces of paper) of the year.

In one of the best pieces of book marketing I have ever seen book wise via YouTube is from Text Publishing, who released a video of one of Australia's most iconic children's show hosts reading an adult picture book called 'Go The F*** To Sleep'. Text appear to have followed every bit of advice I have given to you in this book to the letter. They launched their book via the video below, a simple reading of a very comic adult book. The book is funny by itself, but synopsis and traditional marketing can only go so far and the book more than likely would have had a slow build. But Text went and found further comic fuel to add to the fire. They hired Noni Hazlehurst, the host of the iconic Australian children's show, Playschool. This wonderful host, who gently led children through cheerful songs, and squeaky clean picture books for well over two decades, is reading the book as though to an audience of children, making innocent side comments as she swears at the end of every page.

> If you're interested, check out the video of Noni reading the book here: *https://vimeo.com/26403238*
>
> If you're a Samuel L. Jackson fan, he also read the book for the US publisher: *https://youtu.be/m0jCsKbPSpc* and here: *https://youtu.be/Cb0t9TUNLpg*

Not only is it comedy gold, but a publicity goldmine. All they did was advertise it to their mailing list, and promote it on their social media. In almost 6 hours it got posted 66,258 times on Facebook before it got removed by YouTube for 'harmful content'. But from there it didn't matter, as soon as they placed their video on other video sharing sites fans of the reading quickly passed on the link to the new video. Though Text would have hired Noni you will notice that this major publishing company chose to use all free avenues in placing their video online, and promoting through email and social media. This is because these avenues really work, and if you are still ignoring them as an indie author you may as well go the f*** to sleep.

While this book has already seen great success in the US through Samuel L. Jackson's reading of the book, it's worthwhile noting that different approaches and personalities will work better for some countries then others. Noni is a perfect choice for the Australian version of the book, however the majority of the US and UK wouldn't even know who she was. As a traditional publisher, Text only publishes in one country, Australia, so it works for them. However, as an indie author if you want to reach several countries at once it is worth while playing on common comedic elements and topical subjects.

The above reading is how it's done my good writers; this is how you promote an e-book successfully. So time to add some life to your book. Go forth! Discover your goldmine.

3.13 Advanced Techniques For Generating Demand

We all like to think we are special. It may not necessarily be the case but we continue undeterred. We like to think we have something to bring to the table, something unique, such as our slightly off-colour humour, our beehive hairdo or ability to pull off the underwear-outside-the-pants look. But when every single person in your field is using the exact same mechanism as you – printed words – it is a lot more difficult to prove just how special you are to everyone else. Oh yes, of course you will eventually prove it with your words, but the reader has got to get that far first. In a world where first impressions count, how do you convince people within the time span of two minutes (maximum) that your book offers something no-one else's does? With a digital product you can't slap a piece of fur on the front cover to appeal to the senses and you can't add shimmery gold lettering to dazzle the eyes. It's about features. So the question is, how can you generate demand? Promising to complete a particular humiliating dare and stream it on YouTube if you get over 10,000 downloads MIGHT get you started, however, I'm going to say it's not going to sustain you in the long term. Just a hunch.

To truly succeed in producing an e-book you need to generate demand and make your novel unique. Cory Doctorow is a master of this. Simultaneously, his novel *With A Little Help* can be downloaded as a free e-book, purchased as a $15 print on demand paperback in four different cover designs or as an audio book download or CD. In addition to this, he created 250 limited edition copies of his book each with their own individual features such as special end papers, illustrations, and SD cards with digital specials such as the full text and audio of the novel. Each copy he sells for $275. By making his work special (and determining that there was a significant fan base for his work) he was able not only to generate a significant sum of money, but connect with his readers on a different level creating a fever around his words. Imagine taking that further and having one special edition made that had the only available copy of a certain story. What would that retail for? $2000? $5000? $10,000? The above is particularly important for popular and established authors with large fan bases. By releasing such limited editions you are not only making more money, but you are creating a rarity and a fever around your work.

Authors and even publishers can take this further with the lowering costs of print on demand (POD). Customising books to include a personal greeting from one person to another is a fantastic way of making your book unique (and allowing you to retail it for more). It could be of a personal nature with a message from the gifter to the giftee, allowing that person to record themselves doing a brief personal video that can be placed on a USB that comes with the print book, or it could be of a corporate nature. By corporate I mean you could customise a particular book for a company by including the company's logo and a greeting from them on the front page of the novel (anything from a "Merry Christmas" to a "Hope You Enjoy Your Flight"). With digital e-books you can make a recording of the company's CEO giving their best wishes to those receiving a copy and put in a hyperlink to the video at the start of your book. For a publisher this could be a fantastic marketing ploy if they allow companies to customise a greeting for the start of a highly anticipated novel to give to their clients several days before its release. The company would, of course, pay for the exclusive pleasure.

Why is generating demand important? Other than increasing your profile, it also allows you to charge more money for your novel than if it was a simple e-book.

So, I challenge you: you think your e-book is special… Prove it.

3.14 Would You Promote My Book? Pretty Please?

I've always liked the idea of my own personal recommendation army which spreads word of my work through the cosmos in a supernova of goodwill. But the reality is that, though good will is great and will get you recommendations here and there, it's money that really talks. You want as many people selling/recommending your book as possible and the more incentive you provide, the more people who will be willing to promote your product. Money makes the world go round, hell, it almost costs you for a sneeze these days, and if you offer readers a cut of the profit to promote your novels, they will look after you. Your readers will move heaven and earth to promote you on their site, promote you to their social media groups, on their t-shirts, in their front yards, in song – anyway they can if you just make it worth their while.

To start with I recommend you offer at least 50% of the profits from an assisted sale to your affiliates. Calm down, there's no need for that stunned fish look. The fact of the matter is that your readers have connections and you need to motivate them to actively use those connections to promote your book. Simply saying, "Would you promote my book? Pretty please?", is not a large enough motivator and you are deluding yourself if you think you are going to get more than one brief mention in their Facebook status.

If you are an Amazon or Smashwords author you are already a part of this affiliate cycle. When Amazon or Smashwords authors first publish a book they allow affiliates to send people to their book for small percentage commission if they help make a sale. Those affiliates can be other authors, publishers or just people in the marketplace looking to make some money. Either way, you will sell more e-books this way than you would on your own.

However, if you want to stand out from the crowd, if you want people to rave about you, then you can increase the mullah you offer to boost the enthusiasm of your raving hoard of affiliates. By 'juicing' the affiliate percentage (as Smashwords terms it), you are making your novel a more sought after product to sell. Also, if you can provide your affiliates/readers with marketing material (including Tweets that get a lot of clicks) for the novel, you make this deal a no-brainer. They don't have to come up with the description, they don't have to come up with the Tweets, they don't have to summarise your work so it appears enticing to their friends, they don't have to compose the promotion emails to send out; basically, they don't have to do any work at all. If you provide all the writing I have directed you to do throughout this book as material for them to use in promotion AND you give them 50% of the profits, then it's just a matter of where do they sign?!

There are three very simple steps to getting a good affiliate program going. The first step is to have somewhere that the reader can sign up to be an affiliate. While in the long run it will be better to set up your own affiliate opt in and payment system, this costs money and a bit of professional computer coding. An easier option is to make use of systems already in place, such as Amazon affiliates (*https://affiliate-program.amazon.com/*), Smashwords (*www.smashwords.com*), or Clickbank (*https://www.clickbank.com/*). These sites accept the money for you so you don't have to organise your own payment system. Clickbank is where you can advertise courses or print books that compliment your e-book so that affiliates can easily sign up, get their own personalised link and start promoting. This is probably more helpful for non-fiction authors who are selling their e-books along with enticing extras. The ebook distributors are probably a more viable option for e-books in general. It's important to note that Clickbank does not withhold tax from your affiliate, so it doesn't matter what country they are from, they will still get the full amount. Smashwords however does deduct tax, which could deter non-US affiliates. It's up to you which one you prefer to use.

The second step is to make sure you 'juice' your affiliate percentage if you can to at least 50%.

Step three is to let your readers know that they can actually get a cut of the sale if they recommend/promote your e-book to others. Do this on your website, at the end of your synopsis, on your author page and on the last page of your e-book. You need to provide them with

3.14 Would You Promote My Book? Pretty Please?

a simple list of instructions on how to do this. Give them the link to the affiliate sign up page. Then give them the URL for your e-book or author page; at the bottom of the page will be their affiliate link. Lastly, give them the link to your marketing material and they will be ready to promote you to the cosmos!

But do you know the greatest affiliate marketing ploy? For the first several months, allow affiliates to keep 100% of profit from their sales. That's right; I said to give them **EVERY CENT** to promote your e-book. This is the ultimate viral marketing. It is like giving your book away for free however, rather than waiting for people to find you in the search engines, you have an army of money hungry readers actively finding your target audience and sending them direct to you! There are no other e-books for the new reader to look at, only yours. By telling your reader that once they buy this book they have the right to promote and sell it AND keep all the profits, you will not only increase your rankings in the best sellers list but also recruit soldiers for your recommendation army. Later you can decrease their percentage to a more profitable level for you.

Convincing affiliates to sell your e-book takes advantage of viral marketing, your own social networks and all of your affiliates' networks too! The quicker you get the word out about your e-book, the sooner you will reach the success you are aiming for. All because you went a little further than saying, "Pretty please?"

3.15 Want To Charge Top Dollar For Your E-Book?

Two things make writing exceptionally hard. One is beginnings. How in God's name do you begin? You have the info, you have the idea, you want to entertain but instead you just stare, frowning furiously, at that blinking cursor. Each day of writing this e-book I have frankly dreaded that cursor; it is haunting my dreams. The second thing is pricing. How much are you worth? And how much is reasonable at the other end? We have tested our prices (or you should be currently doing so) to find that optimum point where we make the most money for the number of sales. But our profits are lower than we wanted, so now what? You CAN sell your book for a higher price, but if you want to make a comparable number of sales you must add so much extra value that the price seems not simply unreasonable but so ridiculously low that the reader feels like they are ripping you off!

This ploy will work best for non-fiction authors, children's book authors and already established popular authors. For those of you just starting out in fiction, the previous discussions on short stories and creative interaction with your readers would be the best avenue for you. Also, recording yourself reading your novel as an audio book may be a good way to justify a higher price for your e-book if you offer both as a package. However, if you can think of other value adding strategies for emerging writers I would love to know!

For non-fiction one of the best ways to add value is to interview experts in your field. In this way not only are you adding to the information presented in your book, you also create an alternative way for someone to learn the information. Most experts will let you interview them because it is basically free publicity for within their niche. You shouldn't need to pay them because you are providing them the valuable opportunity of an hour long exposure ending in promotion. If they demand payment, find someone else to interview. They can be leading people in your community or someone with authority in a relevant area.

You may say, "Well that's all well and good Em, but what do I say when the interviewer wants to know what I'm going to do with it? Lying is naughty. I can't lie." Well really there is no need to. An expert should value any opportunity to be promoted, particularly when you will be actively promoting them as an expert in the field and allowing them a chance to give their details at the end of the interview. However, I do suggest that you just tell them the truth: you will turn the interview into a product that sells over the internet, that you will take all the risk by paying out of your own pocket for the promotion and you will produce, edit, market and deliver it. If it makes you feel better, record yourself saying this at the start of the interview and the expert agreeing to it. If push comes to shove, you can allow them to also have rights to the interview. In most cases they will do nothing with it, other than occasionally listen to it for an ego boost and perhaps whip it out for show and tell at their next dinner party.

When interviewing others you must make sure that the quality of their answers is high – don't let them just glide through! If something needs further explanation, butt in and get details or ask for specific examples. The more credible your interviews are, the more credible your e-book will be. The interview needs to be relaxed and conversational. Generally you should provide the expert with a list of questions a couple of days beforehand. Ask open questions to make it easy for the expert to give information packed answers. Listen to the questions asked in the *E-Book Revolution Interview Series (http://www.cravenstories.com/books-and-more/courses/interview-series/)* for examples of open-ended questions. You will find many of the interviews have variations of the same question. The single most important thing you should remember when conducting an interview is to ask your question then shut up! There is nothing more annoying than the interviewer making loud murmurs of agreement into the microphone. There are interviewers I could have calmly strangled with their microphone cord because of the noises I heard from them. Several put me in mind of the sound cows make while chewing grass.

There is no need to get too fancy however; you have written the book so technically you are an expert in your niche too. As such

3.15 Want To Charge Top Dollar For Your E-Book?

you can organise a friend to do a series of interviews with you on the different sections of your e-book. Give extra tips and tell stories that your readers can relate to. Some people learn better through listening rather than reading so this way you are appealing to all the senses. Depending on the number of interviews you include, your e-book could rocket from $2.99 to $30, $50, $90 quite comfortably.

Interviews can be done by anyone. Heck, if I can do them, you can do them. I've conducted interviews during my lunch breaks in the empty underground car park next to my work. I have a mobile broadband dongle for my laptop so I can get internet. Once, I almost ran out of battery 40 minutes into one such interview. I had to run through the car park carrying my laptop, microphone turned off. My expert continued to give their answer oblivious to the fact that I was breathing like a horse after my sprint to the nearest power source. Another time, (though I'm shame-faced to admit) I asked an expert a particularly involved question, just so I had enough time to dash off to the toilet and back again before they finished speaking.

There are several ways you can record an interview. I myself have conducted interviews with people living in the US, the UK (while I was in Australia) and once with an expert in the UK and an expert in another state of Australia at the same time. Skype (*www.skype.com*) is by far the best way to do an interview and now allows you to record your conversations. Skype will allow you to talk with another Skype caller for free over the internet, or you can purchase Skype credit which allows you to call actual phone numbers at amazingly cheap rates. So for example, if a person can only conduct an interview by phone, you can call their landline in the US when you are in Australia and it will cost you about $3 for an hour and a half of talk time. Simply record the call with Skype and you are in business. From there, you can edit out all of the bits you don't want: coughs that sound like you're dying of the plague, sneezes that sounded a little too sticky, those long............ awkward pauses or information you know will ruffle too many feathers. One great (and free!) audio editing software I use is called Audacity (*http://audacity.sourceforge.net*).

Is there a step in your non-fiction e-book that you can actually show your reader how to do? Humans identify much better with visual cues than description. Using software called Camtasia (*www.techsmith.com/camtasia*) you can actually record what you are demonstrating on your computer screen and add that video as another element of your e-book. If you are a travel writer, consider taking a series of short educational videos that focus around your niche; you can have a clip of you eating fried scorpion in Beijing, or one 'guiding' your readers on walking tours around the city you are describing. Homework exercises and home study programs around your books are another way to add value. If you offer enough value in your home study course, along with email and/or phone support for reader questions, you could charge into the hundreds for something that started as a simple e-book.

It's important that you stress the amount of value the reader is getting. By calling your added features 'bonuses' it helps a reader to see that the extras are there for added value. Assign a (fair) monetary value to those bonuses so they can see exactly how valuable the package is. You can chose values based on how much it cost you to collect the information, how much it would cost for the person to hire you personally to teach them or how much it would cost them to hire your interviewed expert for an hour.

To make your offer completely irresistible and the best value e-book your reader has bought, *ever*, you can provide an ironclad guarantee. Give them a complete 30 day money back guarantee, no questions asked (except maybe "How could I improve this?" and "What was your biggest issue?"). This, like a free e-book, completely eliminates the risk for the reader and leaves them no reason not to buy. If you have created a good quality product, very few people will take you up on the refund. This is mainly because of how much quality you have delivered and partially because people are lazy! If you have one in three people asking for their money back, you know your book needs improvement. By asking those customers for their feedback on how you can develop you can build up your book's quality so that only one in 10 asks for a refund, and that's just because they're tight bastards.

Children's book authors can add all kinds of value with extra activities. Pages of illustrations that parents can print out for the children to colour in, cut out activities where they stick things

3.15 Want To Charge Top Dollar For Your E-Book?

together, dress up paper dolls or animals, construct dioramas, join the dots, word finds and even extra illustrations from which the children can create stories. You can add as many extras as you like because you do not have to worry about the cost of printing all these activities; your story is an e-book so the parents who buy the book can print whatever they find most interesting and useful.

For established and popular authors the ways to add value are immense. Not only can you make the audio book available for download with the e-book but you can record yourself reading passages, include interviews with other major authors you know in the field, create a home study course teaching people how to write for a certain audience or genre, conduct Q & A sessions through webinar, make copies of your public appearances available for download and add bonus short stories, novellas and 'deleted scenes' that didn't make the edit. The amount of value you can add is only limited by your imagination. Applying Cory Doctorow's method of creating limited edition copies of your work is another avenue you can pursue.

While endings are seldom easier than beginnings, if you want your (happy) ending to include selling your e-book for an above average price, you have to make the experience worth more than the money paid by the reader. These extras only have to be created once, and if it brings in the extra money you want then surely it's worth the effort.

3.16 Ten Shades of Author Collaboration

Why is it that every time I talk to writers 'new' to the digital world I end up feeling like I've volunteered to be a crash-test dummy for a dozen cars? For a while it happened every time I came home from the latest writers festival I attended. Even after all the studies, the blog posts, the infographs and pie graphs, authors and publishers are still under the impression that only the few elite will ever be writers and apparently the whole population of worldwide readers will only read the books of their one perfect author. That god awful vibe still exists, insisting we (authors) are all in competition with each other, and if we even get out readers to look sideways at another author's book, we will lose their hard won attention, *forever*!

"Let's face it, not everyone can be a writer, there's not enough room," was the mantra I heard again and again. I'm surprised I didn't see more writers wandering around the festival looking like escaped kidnapping victims with hoods duct-taped over their heads. Setting aside the fact that I've been asked to swallow a steaming pile of BS, let's break this down a little. There a millions of readers across the globe, it is ridiculous to assume they will only read one book or one author in their life time (40+ reading years). Let's assume only half or a quarter of the millions were voracious readers, where reading (and I know a lot of you feel me here) is a way of life and once you've made it through all those Harry Potter novels you need more pumpkin juice.

I can name a dozen favourite authors off the top of my head, none of which are producing books fast enough, *collectively*, to fill more than two-three months of my allocated reading time. So what do I do for the other nine months (Or eight if you subtract a month's worth of social media)? Making floral arrangements? Completing my ninja training? No! I'm sampling the wears of new authors, who may not have as much 'talent' as my favourites but are competent, and entertaining and still worth my reading time. They may not be the elite writing-del-a-creame, on the NY Times bestsellers, but they write books in genres I like to read.

While readers in the past were restricted to by the cost of printed books, with digital books people are finding reading more affordable. Surveys as far back as 2010 were already noting that 40% of people using e-readers are reading *more* books now than they did when they were reading print books and that number has only increased. In America surveys suggest that 31% of the population own tablets and 26% own e-readers.

E-book authors are in a unique position where we have a sector of the market that's at least *50%* voracious readers, ready to snap up a good recommendation from an author that they just read and loved. So why wouldn't you, as an author, collaborate with other authors whose work you enjoyed, and send them your readers? Think of this as a marketing exercise: if a reader is done with your books (yes, all of them), how do you continue to build their trust in you? By giving them good recommendations, and in the process, another author is building the trust with their readers by sending them to you. It's a simple concept. No two authors follow the same marketing strategy, which means that every author will have reached different people. No two sets of fandom are the same.

At the risk of sounding cheesy, we need to share the love, not horde the treasure. Because let's face it, everyone dies, and hording treasure never to use it, is stupid. Better to make connections and friends and promote an abundance economy (Channelling Seth Godin's Icarus Deception here). As the ever wise J.A. Konrath says:

"One hand should always be reaching up for your next goal. The other should be reaching down to help others get where you're at. We're all in the same boat. Start passing out oars."

So what are authors doing to promote not only their own work, but the work of their collaborators? One of the best examples I've come across is the hugely successful *Ten Shades of Sexy*, a permafree ebook which features sex scenes from ten different novels and authors. The premise is simple; ten authors got together to produce a compilation book that brought a reader nothing but the 'good parts' of their sexy romance novels. They knew their audience well, and took advantage of the knowledge that romance readers loved to bookmark 'the good parts' of their favourite novels and gave them just what they wanted. The brilliance of this project is twofold; you have ten authors promoting the *same* book to their networks of *different* readers. By supporting each other's work they are building their own fan bases. Secondly, the book is comprised of novel samples, so the ultimate aim is to use this free book to urge readers to purchase the full novels.

At first glance looking at Goodreads reviews you would think the exercise had failed with an average star rating of 2.98. But if you delve into the reviews you find that the reviewers have given it 3 stars because the 'novel' doesn't stand as its own book (remember, this was not its purpose – to read like a rounded anthology). The reviews then go on to admit that at least every reader put one or more of the novels sampled on their to-read list. That *was* the aim, to gain more readers and sell more books; *Ten Shade's Of Sexy* did what it was supposed to do. I will admit I read it, and love it for a different reason. I got to read the racy parts without having to go through what I saw as the tiresome parts of a romance! But that's my genre preference. It's worth noting that this book was still in the top 100 free Amazon Kindle downloads after almost 8 months after its release.

While this is a great example, this may not be the ideal way for authors to collaborate in other genres. I have blogged about this before but consider going out right now and reading other indie authors in your genre. If you like what you read, and feel your readers would enjoy it too, approach the author about collaborating! Put the synopsis for their novel at the back of yours and have them do the same for you. Then every reader who finishes your novel, full to the brim with praise for your prose, sees the ad and because they liked your book so much they take your recommendation and purchase the other novel. It's a win-win situation for both authors.

You can be very clinical about your choice of collaborator if you wish, asking them about the size of their fan base and the number of downloads they have, making sure they are similar to your own. You could also wear a Dracula mask and shout *boo* at them through their living room window, but hey, if you want to scare off collaborators it's your choice. You can even join the affiliate programs of Amazon, Smashwords and Kobo and get paid a percentage for every collaborator novel you sell through the back of the book (how I would do it).

So let's collaborate with each other, spread the love and ignore the treasure hoarders.

3.17 Webinars: Connect With Readers Across The Globe & Generate Massive Sales

There are many things you can do in your pyjamas: sleep... write, read, make dinner, eat dinner, have a party with a bunch of eight year olds, take out the garbage, go to the local store for a snack, impersonate John Lennon, sleep... It had never occurred to me that you could sell anything in your pyjamas (other than more pyjamas), especially not books. Ten years ago, connecting with someone on the other side of the world in real time, without a ten year lag time between sentences, was only dreamed about. But a lot has changed in ten years (not in the least pyjama fashions) and one particular piece of technology has got me hotter under the collar than a paedophile at a Wiggles concert.

There has never been a greater time in history for people to personally connect with someone on the other side of the globe. Your readers could be at a cafe in a busy street, on a beach in the Caribbean, or sitting in front of the computer wearing their girlfriend's underwear; as long as they have an internet connection, they can communicate with anyone in the world. Never before have authors been so accessible, yet very few of us are taking advantage of this.

Traditionally, connecting with your readers meant you had to be in the same geographic location as them, either on a book tour or other organised writing event. You smiled until your face hurt, signed a couple of books and forced yourself to laugh through badly delivered jokes. And, while this e-book has given a myriad of strategies on how to create a reader experience through words and videos and recordings, these methods are in some small way still distanced from the reader. They are generic, there for anyone to find. Words take on different meanings when you cannot hear the tone or see the facial expression of the person. This is unfortunate because, as authors, most of our convincing is done through the written word. And as such the powerful emotions your novel synopsis is supposed to invoke can be lost in translation, and sales lost along with them. In this digital age we have the power to interact with readers personally, so why should we settle for the out dated way of doing things?

Enter the Webinar, your rocket powered jetpack to the other side of the world. Connecting 'face to face' with your readers is still the best way to sell a book. A website or a synopsis is almost like trying to cold call your readers; they don't know who you are, and if your wording doesn't get the correct meaning across it's goodbye sweet cheeks. A webinar gives you the opportunity to have this face to face interaction with multiple people around the world at the same time! Basically, a webinar is a web based seminar which allows people to hear you and at the same time you can either be transmitting a PowerPoint or a live screen demonstration directly from your computer screen to theirs. Listeners can type in questions and make comments, allowing you to connect with them personally without leaving the comfort of your home. Readers love this interaction and being able to hear the writer speak to them directly.

A webinar sales presentation reaches out and talks directly to the audience, entertains them, answers their questions and demonstrates the presenter's personality and style, unlike websites or e-book stores, which are impersonal and static. You can do webinars from home, they cost you almost nothing to set up, you can make money with them easily and they work in *any* market. GoToWebinar is one of the main platforms for conducting a webinar (*www.gotowebinar.com*). A free one you would probably know well is Google Hangouts (*https://hangouts.google.com/*).

For fiction authors you can run Q&A sessions, talk about writing in your genre, and even teach about it, or do readings of your newest novel. At the end of the webinar if you wish you can sell your latest book to the listeners for a one day only special webinar discount price. If you are an already established or popular author you can use a webinar to launch your e-book the day before it is available in bookstores. By simply reading out passages from the yet to be released book you can whip readers into a frenzy.

For non-fiction authors you can take the webinar, and your profits, so much further. Not only can you charge a higher price for your novel in a webinar, but webinars have an almost 10% conversion

4.17 Webinars: Connect With Readers Across The Globe

rate in comparison to a website, which only converts 1% of its visitors. As I noted in an earlier chapter, you can either sell e-book singly, with a set of interviews or a home study course or even with 12 weeks of personalised coaching from you (which you can conduct through a webinar). By giving people multiple ways to learn, including and beyond reading your text, you can create a whole new reader experience.

There is a fine art to selling a non-fiction e-book and any associated experiences. As with all of the marketing I have talked about, you cannot make your webinar into an hour long sales pitch. Droning on about how good your product is will have your potential readers throwing in the towel in disgust and the webinar will be empty before you're even half way through. Then you are left talking to yourself, and that's just a little sad. You need to have your ultimate goal in mind and structure your presentation so it leads naturally into your offer at the end. Firstly you need deliver a big promise: what are your listeners going to learn? What is the benefit for them? Remember, if you tell them you are going to teach them something, you'd better do it! Or you will have as much chance at making a sale as you have of finding a golden ticket in a Wonka Bar.

Having made your opening promise, introduce yourself and tell people how you got to the point you are now. Let them relate to the emotions that led you to seek out and master this information. Provide some statistics on why this information is so important and give general evidence on the trend.

Then you need to deliver good solid content and information. The best way to teach it is in steps (three, five, or seven) so you can break it down into a sequence people can easily follow. You really need to deliver your butt off here, and give as much information as you can in 40 minutes. Be the Brangelina of the webinar world – make them swoon! There is only so much you can talk about in 40 minutes and it will barely touch the tip of the iceberg where your e-book is concerned, so don't feel that 'giving away' your tips will lose you any sales. The greater the quality of information, the greater your credibility and the more people you will convince to go further with you. There is no need to be nervous when giving your presentation.

One reader's experience varies greatly from the next but the best way to deliver a good talk to all is by pretending you are talking to 12 year olds. No one will ever complain you made it too simple to understand. Use stories to help illustrate your points rather than just instructing. Draw the listeners in. At the conclusion of your information section summarise the steps. Remember, people are coming home from work, school or university and they are choosing to spend their time with you instead of watching TV, making out with their girl/boyfriend or any of the other million and one things they could be doing. Be entertaining and worth those precious minutes.

There will always be at least 10% of crowd who want to go further and faster with what you have. At this point, let them know that if they would like to learn more you have an e-book or a course etc. You need to make this offer irresistible. You need to give more value than you are getting in return. Offer something unique, whether it's personal attention, a weekly critique of their writing or just something that your competitors aren't offering. Or even just something you can do better. As always a guarantee is great as it eliminates the perceived risk to the buyer, making them more inclined to purchase your product. This is similar to how you would give a book away for free to attract readers, however this way you see some immediate money, and if you have a worthwhile e-book it is unlikely you will have to refund anything.

Though writing is a creative outlet we must realise that, if we wish to make money, it is also a business. Webinars are a fantastic way to make your business unique. Just by being accessible and offering an extra service that other people don't, you are automatically ahead of your competitors. You can beat them with the strength of your little finger alone, because readers want to connect. And people will give you money if you give them what they want.

***

Creative Ways To Interact With Your Audience

A webinar is just one way you can creatively interact with audiences. In this digital age, if you are someone like Cecelia Ahern, backed by a large multinational publisher like HarperCollins, you can place a

4.17 Webinars: Connect With Readers Across The Globe

specialised glyph (or picture/character) within the pages of your book. When your reader goes to the publisher's unique author website, they show the glyph to the webcam and get instant access to a whole new virtual world inspired by your stories.

While many indie authors can only dream of having such a creative interaction with their audience, there are still achievable ways you can foster your author/reader connection. Why wait for publishers to lead the way? In this digital age, abounding with free technologies, we are able to forge the new frontiers ourselves. As mentioned before, you can upload your own video trailers to YouTube and even encourage readers to upload videos of themselves acting out the scenes (whether it be for a competition or just for fun). If you have the patience to do a bit of extra formatting you can even add illustrations to your work. Stick figures are always popular…

Another fantastic way to interact with your audience is through podcast. Scott Sigler is the king of this particular method. Scott created the first podcast only novel in 2005. Today, by giving away his self-recorded audio books as free, serialised podcasts he has gathered a massive following for his work. A large slab of his following was formed BEFORE he was traditionally published. You can think of a podcast as a more modern version of the early serialised radio fiction, without the fake British accents. By serialising the e-book not only can you gather a following but later, when it is complete, you can sell it or give it as a bonus with your e-book (to entice people to buy and also to justify a higher priced book).

Subscription websites (for example *http://www.astorybeforebed.com*) also offer a great chance for indie authors of children's books. These sites allow a parent to record a video of themselves reading a children's book (selected from the list) on their webcam. This recording can then be played by a carer for the child while their parent is away. This allows your reader to add a personal touch to your work.

The more places readers can find your work, the better!

3.18 A Writer's Money Isn't Just In The Books

It's not many a writer who would admit that comic book companies are the geniuses of our industry. No seriously. They have the whole demand scale figured out. Not only do they mass produce paperback copies of their stories but they have television shows, movies, yearly conventions in every major city in the world, they have lunch boxes. They have toy figurines of the hero, the side kick, the villain, the villain's hairless cat, and let's not forget the sidekick's landlady. And depending on when they're made, how rare they are, and whether or not the buyer has resisted temptation and left the figure in its original packaging, the villain's hairless cat can go for several hundred dollars when first sold and several thousand dollars years later. This my friends is marketing genius, realising that the money is not in the paper bound book, but in the other entertainment opportunities we can provide the audience based on the story.

Indie publisher Richard Nash talks most eloquently on writers needing to expand their scope from the novel to further interactive opportunities like workshops, Q&A sessions, memorabilia, exclusive dinner parties, your own board game or selection of swim wear (well you never know) and endless other possible endeavours depending on your genre. Larry Correia, author of the Monster Hunter International Series, when he first started indie publishing he encouraged the design of military style patches for various teams in his series. He also actively encouraged people to buy not only signed books from him but patches of his own design as well. A German art student, Benjamin Harff, made a beautiful hand-illuminated and bound copy of J.R.R. Tolkien's Silmarillion (*http://tiny.cc/TolkienArt*), an enhanced version of the book that many Tolkien fans would give their pet orc for.

Slowly authors are coming to the realisation that they can create a fever around their work by allowing it to move outside the written word. One such author is Garth Nix, a world renowned fantasy author with books published in Australia, US, UK and a dozen others. Like many well-known authors, he could have just stuck to his paperbacks. But clearly Garth is also a savvy business person and saw the opportunity to deliver something more to his fan base. Garth is best known for his Abhorsen (or Old Kingdom) Trilogy and he leveraged the books' popularity to create another sort after product.

"In these books there are necromancers who raise and control the Dead using seven named bells. These evil necromancers and Dead themselves are opposed by a family called the Abhorsens who use their own versions of the bells to make sure the Dead stay in Death and do not trespass into life. I thought it would be great to have silver charm versions of these seven bells."

Garth created Stirling silver bell charms for a charm bracelet based on the core idea in his fantasy series. Each 'bell' has its own individual mark, for each bell has its own name, and you could choose from not one, but three different finishes. 'Bright' which is brightly polished; "Ancient" which is a duller finish and 'Black Handle' *for those evil necromancer types.* The charms ranged from $39.95-$45.95 each. Garth further increased the rarity (and hence the value) of these charms by identifying whether the charm was done in the first casting, in the second casting etc making the first casting the most valuable in later years. Unintentionally (but effectively), Garth built up the excitement of his readers by announcing his intention almost a year before the charms were finally ready. Furthermore, within the website he refered people to buy his books if they want more information about the bells and their uses, ensuring further sales of his novels as well.

So if comic book publishers are selling action figures (or very expensive evil cats), illustrators are making works of art and fantasy authors are forging bells out of silver, what could you create from your words?

3.19 The Business of Being a Writer

I have a notoriously faulty memory, to the point where I will call family members on their birthday, not to congratulate them but to talk about some mundane subject I had been meaning to call about. Diaries rarely work, normally they end in some hidey hole I don't find until I move house. So I have taken to writing things on my hand with pen in a bid to appear organised, and on occasion, prompt my memory. Today's reminder came in the form of **BC**.

After rejecting several meanings, both religious and otherwise, I finally remembered what it stood for: Business Card. Writers rarely think of what they do as a business, that every person they meet, from the pretty girl behind the counter of the cupcake store, to the editor who gives them 2 mins to pitch their novel, is a potential champion of their book, their information and a possible link to media coverage. For indie authors this can be vital, because most people will keep a business card over anything else given to them. They are terrific promotional tools because they are inexpensive. They are also terrific icebreakers and informational aids.

A business card has to have a specific purpose, and as a published/self-published writer I would recommend you have two. One specifically for your e-book or book, and the other for you as a writer and (if writing non-fiction) an expert. My business card's purpose was to establish myself as an expert writer and e-book specialist and to recruit Joint Venture partners to help me sell my e-book. You will remember from my previous posts, that a Joint Venture partner is part of your army of promoters. They do not promote you to get that pleasant fuzzy feeling inside; they will promote you in exchange for you giving them part of the profits. If you give them money to help they will move heaven and earth to promote you through social media, on t-shirts, in song... You take advantage of the people they know, to become more widely read and recognised.

The result of my efforts is below (though I should say this was my very first business card from 5 years ago, I've morphed and changed things up since then, moving websites and branding as I found my groove):

On the front of the business card I had the cover of my e-book – simple and eye catching, done by a professional cover designer, not my next door neighbour's two year old. I had the name of my e-book and a six word pitch or 'silver-bullet' that tells people exactly what the book is about. It is short and concise and does not include the entire storyline of my novel or dramatic and unnecessary words like 'stupendous' and 'so funny it will make your socks fall off!'. It gives my name and expertise along with my phone number (more relevant if you are recruiting your promotion army in your own country), email, blog and the webpage where I first sold my e-book. The colours of my information matched the colours of my book cover. These days, particularly if you're going for a writing career, I'd put a photo of you holding some of your books (such as the photo on the next page) so you don't need to change up your business card too often!

3.19 The Business of Being a Writer 153

There are ads all the time on ebay promoting 250 free business cards and all you have to pay is postage. So no more excuses! While your e-book may be digital, your promotion should not be restricted to the screen. There are so many networking opportunities out there, conventions, festivals, talks and courses that you cannot afford to hand people your information on a piece of scrap paper.

3.20 Why Give Your E-Book Away For Free?

Free can be a very tricky thing. One cannot deny that free is fabulous for the receiver because they have no risk. It's like Christmas without having to give thought to what you should give back, being invited to have tea with the Queen without having to perform a single act of bravery or aging to the point of spontaneous decomposition. For an indie author, giving your work away for free is about the only way you can guarantee a glance at and perhaps a curious download of your novel which otherwise would be an Arthurian legend. You know, a completely immoveable sword stuck fast in stone until the golden hand of Arthur (or luck) comes along. In the writing world this golden hand of luck is also known as the occasional reader who will pass their enjoyment of your novel onto their friends.

Why would you give away something for free, on the off chance that the reader will remember to tell their friends, their relatives and their pet goldfish? This question has stopped many an author in their promoting tracks. People are forgetful creatures, and if they are like me, they download the free copy (just so they don't forget the name), add it to the end of the 20 book long queue (which constantly gets updated with new books from favourite authors) and by the time they get to reading it, telling their friends in a visible and public way is the last thing on their mind. Hence authors get caught in this Catch 22. Free is how you move forward, but people are notoriously lazy and absent minded, and unwilling to pay for an unknown book from an unknown author.

But wouldn't you know it, there are some fantastic strategies and tools that allow you to provide a free copy of something *and* get promotion for your book in one neat package without 500 downloaders nicking off with your work, never to return. I'll talk about two of them here.

The Social Media Approach

The first tool I want to talk about is a website called Pay With A Tweet (*www.paywithatweet.com*). Basically you can create a button for your webpage, Facebook page, blog, or author profile that allows people to download your product (and in our case, e-book) for the price of a Tweet. The reader can change the Tweet to whatever they wish (from your cool, entertaining, carefully worded one), but they cannot change the link that leads to that Pay With A Tweet button. Therefore, to get the free novel they have to Tweet about it first.

The power of the Twitter phenomenon has barely been tapped. Twitter is used by every well known company, department and personality in the world. NASA has Twitter feeds for several of their projects and Tweets have even come from outer space. The LA Fire Department spread information about the 2007 fires via Twitter because the news channels weren't fast enough, and that could mean a life saved or burned. The general public see social media as a way to 'keep the bastards honest' as we say in Australia, meaning keeping big companies honest on everything from the quality of their service to offers of a free product. As such a Tweet or any other social status update from a friend is seen as a true testimonial of a product or producer to their trusted circle. Pay With A Tweet can help authors in an unprecedented way, allowing them to take advantage of the 'free' phenomenon and get a guaranteed promotion at the same time.

Imagine the possibilities! This is how you integrate readers in such a way that they become part of your campaign. It is a tool people will use over and over again and willingly tell their friends about in exchange for what you offer – namely, their entertainment. One press is currently using the Pay With A Tweet website in a great way. Keyhole Press has several books for sale at an average price of $12.99 OR you can pay with a tweet. What is a person going to pay? The money, or get $13 of value for the price of a measly tweet? I can readily believe that readers will promote before they read without a thought, because they don't have to pay for it. Then the viral marketing truly begins. Keyhole provides no synopsis and no genre, just a brief, ten page PDF excerpt of the file – just enough to give the reader adequate confidence to go ahead and Tweet for the download.

3.20 Why Give Your E-Book Away For Free?

You could use your reader email list as a promotional tool by notifying them of a free download, such as a short story, or if you're a children's book author free colouring in pages, and then have them tweet about it to get the freebie. The readers on your list know your work and will readily promote an author they already know.

From there you can easily track the URL to see how many people have Tweeted your book to the world, all because you gave a reader something for free. You can track the number of people who click on the link by compressing the link to your page (which has the Pay With a Tweet button) using *http://bit.ly* or *http://tiny.cc*. Bit.ly will then tell you how many people have clicked on that link.

The Mailing List Approach

While the first tool takes advantage of the social media approach, another great way to leverage free is through tools like Book Funnel (*https://bookfunnel.com/*). Book Funnel allows you to give away your 'reader magnet', aka your freebie, in exchange for a reader's email address. Why is this helpful when you could put an email capture form on your own website yourself? While I'm the first person to look at the world with a sunshine and rainbows mentality, I also know that getting people to visit your website can be *hard*. So you need to go where the readers are already congregating. Book Funnel is one of those places that readers know they can go to discover new authors (and sign up to their mailing lists in return!).

Book Funnel also leverages the power of author collaboration. The website allows you to join promotions that different Book Funnel authors are holding. They might be a Valentines Day promo, a Christmas Promo, an Australia promo, or a Sexy July promo. But no matter the genre, there is always some sort of promotion running that you can join. You can even be the lead in one. The great thing about these promotions is that each author who joins into that freebie promotion markets it to the readers they have on their email lists and social media, meaning your book freebie or short story extravaganza, could be put in front of hundreds to thousands of people.

While the temptation is there to constantly join a promotion I would caution you against joining more than three or four a year.

If you're advertising how people can get your book for free 24/7, potential readers become as blind to your posts as a bat without sonar. Everything in moderation; impatience is no ones best friend. You want to make sure, after all, that your social posts and newsletters are only 30% promotion, 70% entertainment/conversation.

Book Funnel provides a host of other great features too: ways to securely send ARCs to readers; codes you can sell or give away in-person that allow readers at events to download your novel; and they are even able to help you sell your books directly from your website and deliver them to readers in a way that makes sure they can't sneakily pass the download link onto all their friends.

Instead of paying with money, your readers are paying with social or email advertising. And you didn't even have to buy them a drink to introduce yourself.

3.21 The Ultimate E-Book Launch

I have been told that the biggest highlight in an author's career is their first book launch. What says 'I made it!' better than a party? It also provides a bit of promotion, some good snapshots (sober and otherwise) and you may even sell a couple of books. But if you're an indie author (or even an established author) releasing an e-book, how do you get the word out? How do you launch that sweat soaked manuscript you've been slaving over without being able to entice an audience with a wine or two?

Why, you make it an *event*.

If you haven't heard of Isobelle Carmody's launch site for her e-book Greylands, then it's a little too late as it self-destructed in August 2012. Since the end of the launch the site/debate was archived to *http://tiny.cc/greylands* and *http://tiny.cc/greylands2*. This folks, is how it's done. Isobelle took the book launch one step further and made it into a month long forum complete with special guests, book trailers, and prizes.

Isobelle is a world-renowned fantasy author of over thirty books for adults and children. After taking back the rights for her out-of-print book, Greylands, she decided to re-release the novel in e-book format. It was clear this needed a bit of a fanfare and what better way to do that then to have a debate about the publishing industry's move into digital? I can already hear you weary Indies crying, "Of course this is working for her. She knows people, she has fans, she can call on big name authors because they're friends." Being an established author is of course very helpful, but it is not what makes this launch so successful.

Everything Isobelle has incorporated can easily be replicated by indie authors to great effect. She used the forum to build up the hype for the book several weeks before it was released. In those weeks she announced the prizes (kindles, print books and audio books) for the best comments on the forum, and released a trailer for the book complete with addition video on the making of the trailer. Creating anticipation is key, whether you're newly published, or an established author whose fans have been fighting over out-of-print copies of your books at their local library. On the day of the e-book's release she had another fantasy author 'launch' the book for her in an online speech, again a brilliant move by getting other people involved and promoting her work.

The major stroke of genius in this launch however, was conducting a month long debate over a contentious subject, in this case the digital vs. print debate. There are many contentious subjects out there *cough *cough *religion* *cough *cough *politics* *cough *cough *guns *cough *cough, that an indie author can tap into to create a bit of a buzz around their release. Every day this month a different person has posted their perspective of the topic on the site, taking the debate in a new direction and allowing the general public to join in. While Isobelle did use her connections to well know authors, she also invited opinions from journalists, editors, librarians, readers, teachers, professors, emerging authors such as myself, and high school kids who will be the generation leading this revolution. See my post below:

So perhaps it's time to start harassing your contacts on Facebook and include an e-book launch in your marketing plan. Every book has an angle it can launch a debate around; what's yours?

3.22 How To Get Your Books Into Your Local Book Store

In any industry focusing on only one aspect can go one of two ways: you become the 'expert' and dominated your niche empire astride the backs of two snow white polar bears; or you crash and explode, taking all your hopes and dreams to a fiery death because you didn't diversify and are now eating homemade cardboard cereal . I'll let you decide which of those two options is true for most people (let me give you a hint, polar bears are now endangered).

While the Big 6 publishers are finding their sales flagging due to digital (somehow they don't seem all that receptive to diversifying in a reasonable e-books-do-not-equal-$10 way), indies on the flip side are finding that focusing on digital is halving their potential audience pool. As much as digital publishing has made massive leaps and bounds over the past five years, print still takes up 70% of the market (or more in some countries). As an indie author you want to be able not only to satisfy your own ego by holding a book in your hand, you want to reach the 70% of the world who like to cause themselves shoulder troubles by carrying around books in their bag.

There's only one problem, while you are clearly better at the digital stuff, traditional publishers are better at the print and they have all this distribution-of-a-book-in-the-real-world stuff figured out.

Which begs the question: once I have print copies of my book, how the hell do I get them into stores? Well, this post will help.

The reason I decided to put my own books into print was three fold, and unsurprisingly the reason was different for each of my books:

- For my non-fiction book, *E-book Revolution* (teaching writers to create their own e-books), I realised that most of the authors I was teaching had never read an e-book in their life. So trying to send them to resource that was only available as an e-book back fired like a man who's eaten too many beans. Even though the irony killed me, having a paperback version of *E-book Revolution* made sense.

- For my YA comedy, *The Grand Adventures of Madeline Cain* (A novel formatted as though you are reading the main character's Facebook page), I very quickly realised that my target audience couldn't legally own credit cards which made the purchasing of digital books a bit more of a barrier then I originally thought. If I wanted to do talks in schools, or at youth festivals, or get the book into the hands of young readers at all, I had to give them something they could buy with old fashioned pocket money.
- For my short story and play, *Jake's Page* (which brought Facebook to the stage), it was a matter of how directors and actors work; by highlighting, cutting things up and drawing all over pages! A print copy was going to allow this a hell of a lot better than digital.

So I spent months pulling my print on demand copies together (you can read more about that journey here) and at the end of last year was holding shinny new copies of my books! Oh the smell! *snort* Then the hard work of getting them out into the world began.

It was amazing the sort of attention I started getting from my friends and family when I began inviting them to signings and posting pictures of those signings on Facebook. Friends, who I love and adore but frankly paid about as much attention to my posts about my e-books as they did to brushing their teeth, suddenly perked up and were asking where they could get a copy. I started being talked about between friends, in writing circles and the next thing I knew I was being invited to nights where teachers and librarian's were speed dating me. Yep, that happened. And from there I even had a very swanky local book store offer to host my next book launch.

My print book versions got me noticed in a way I didn't expect, and now, without a doubt I will be doing any future releases in dual format: digital *and* print. My books are only available in physical stores in Brisbane at the moment (and online retailers like Book Depository) but every few months I expand a little further, and more people in the right places find out about me.

So below are my seven steps to connecting with your own local book stores

3.22 How To Get Your Books Into Your Local Book Store

Step 1: Set Yourself Up As A Professional

The way to stand out from all the self-published authors who stumble into a book store waving their novel about before launching into a monologue about the reason why their book is set in the planet Disy*dhukt!gh, is by being as professional as any distributor or book seller who walks in that store.

That means producing a quality product that looks like it belongs on their book shelves. That means making sure you have your own stock and understanding that they will not buy your books outright, they will take your books on a commission basis. To be professional you need the following:

- Dress up and look nice! Think of this as a job interview, you are trying to convince them to take traditionally published books off the shelf for yours; you need to provide a better first impression then your book cover.
- To be set up or registered as a business in your country. I am registered as a 'sole trader' and have an ABN (Australian Business Number) which was free to apply for. Other independent authors I know in Brisbane have created an imprint for their books and have registered as a small business. You can go for either option. In Australia this Business number is connected to your tax number (or ID in the US/Canada??). The book stores need this number to be compliant in their own tax records. Sorry guys, can't do the whole mafia money under the table trick... If anyone knows what the equivalent is in other countries please drop a comment below!
- Make sure you have contact information ready to hand them (which includes your ABN).
- Know your exact genre, target audience (age), and two authors whose work is similar. It is those similar authors that will tell the manager where you book sits on the shelf.
- Have a short 20 second pitch memorised and ready to go. If you have any experience that relates to writing or your book, have a short spiel memorised about that too. Writing competitions or short story publications aren't normally significant enough to add any weight to your pitch (unless

the competition is one held in high esteem or you are a freelance journalist).

Step 2: Know Your Numbers

You need to know how booksellers work in your country. In general, they will take on your book on a commission basis meaning you don't get paid until they sell a copy. The commission booksellers take can vary. The majority of the book stores I deal with work on a 40-45% commission on Retail Price (though I had one lovely book store tell me they only wanted 30%). Note that this is higher than the commission they get from traditionally published books, because they are taking a 'risk' putting your book up on their highly coveted shelves. Know what the number *should* be, then ask the manager you are talking to what their commission percentage is. If they are way out of left field, negotiate by telling them the commission taken by other stores. They will tango, or they won't, and you'll need to decide if what they are asking is worth it to you. Most stores will take your books for a six month trial period, and if they don't sell, they will return them.

Then you need to make sure your paperwork checks out! Prepare an invoice which includes your business number, contact information, and book information, as well as any tax that is applicable. In Australia, booksellers like you to account for GST (general services tax) even if you are not registered as a business who has to pay GST (you need to earn over $75,000 a year in Australia before you have to register for GST). For the US or Canada because the tax is added on at the point of sale you don't normally need to include it in your invoice unless you're registered for the equivalent of GST.

Don't hand the invoice to them until they have told you they want your books. If they accept (you go girl/man/leprechaun!) you will need to hand them an empty invoice first and email the filled out one later (see below). Each new store should be given a new customer number.

Step 3: Promotion Material

Part of looking professional is coming with your own arsenal of promotional material to blow potential readers out of the water! The more you plaster your surroundings at book signings with things that look professional, the more legitimate you will seem. I have four sets of promotional material:

- *Posters:* They brighten up a signing area and also give the booksellers something to hang on their walls when you're not there. Or they may just keep them in the back until you back for your next signing. That's cool, but at least they know you are serious about promoting yourself well.
- *A Sign:* explicitly saying this is a book signing with your picture and some reviews. It took me two book signings before I cottoned onto the fact that people thought I was saying hello to them because I worked for the book store. Sometimes people just need facts shouted at them with a speaker phone. You can never be too obvious with promotional signs.
- *Bookmarks:* One of the best ways to entice readers, as we all know, is with something free. It's a good way to get people to start talking to you. I made sure that my bookmark wasn't just a promotional flyer, I wanted to front to be something funny that would make them want to use it rather than throw it away. Better they're using *my* bookmark to hold their page in a best seller then have them forget about me.
- *A You-Cannot-Miss-This Banner:* The banner, though a little pricey, has been invaluable as a way to draw attention and brand the front of a bookstore with my own brand! Nothing says pay-attention like a banner that is taller then you are. Trust me.

Step 4: How Do You Approach Your Local Store?

It's always never wracking pitching yourself, particularly to very sales orientated managers who have their judgey faces on. So, rather than start on the defensive, I tend to put the onus back on them with the following:

Me: (*Approaching sales assistant with sweaty palms and erratic breathing, eyes may look a little wild*) Hi there, I'm a local author.

I was wondering if I could speak to your manager for a couple of minutes?

SA: I'll see if they're available.

Me: Thank you, I'll only take two minutes of their time. What is their name?

SA: Blah blah.

Me: (*Manager approaches, inside my brain and stomach are having an argument and throwing chairs at each other*) Hi there blah blah, my name is Emily Craven, I was wondering if you support local authors?

BB: (*Can't say no to that!*) Of course.

Me: Oh wonderful. This is my book XXX, it's about blah blah blah and is for ages blah and sits in the same section as Author McSexy's book MMM. Would you be willing to stock a few copies? I am more than happy to do book signings, I have my own promotional material, and this is my writing experience and amazing story of awesome.

Make sure you come during the work week so you get the main manager. If a manager is not available politely ask how you might set up a time to come in for a chat. If they decline to take books by independents, thank them for their time and then make face in the window at them when their back is turned....

Step 5: How To Set Up A Book Signing & When

When a manager first stocks your books, ask them in that same conversation if you can book a Saturday to do a book signing. Get them to commit then, the sooner you can prove your books sell, the better. Here's how to make sure your book signing is the best it can be:

3.22 How To Get Your Books Into Your Local Book Store

- Saturdays are bookseller's busiest times, that's when you want to be signing.
- Sign ALL DAY. Only being there for three hours will not give you the sales you need to make it worth your while. I go from 10am to 4pm (or 3pm depending on how busy the area is) on the day of my book signings. Make sure the book store is aware of that!
- Start your book signing runs in the second half of the year. For some reason January – June is a book signing dead zone. Try to book into a book store in the 6 weeks leading up to Christmas, that will be your most successful time.
- Make sure you have enough books to fill the signing table, it looks a little sad if there is only a small pile. You will have to bring your own books to sign, and then the reader pays at the till.
- Keep track of how many people buy your book; you'll need this to fill out your invoice later.
- Do more than one signing! The more signings you do, the more books you sell, the more likely a bookseller will be to keep you on the shelf, even if your books aren't selling without you there to connect personally with the reader.
- Say hello to everyone who passes your table! You will not get people coming up to you by sitting behind the table and waiting for them to approach. Stand up! Offer them a free book mark, ask them about their day, and don't be a jerk. The reason I say don't be a jerk is because…

Step 6: What To Expect

…Most people will ignore you when you say hello. It can get very depressing to stand up for over five hours and only have ten percent respond and then only twenty people or less actually approach your table so you can tell them more. It can make you want to be a bit more invasive, to start yelling tag lines at them 'Do you like Facebook?! Do you like Action Adventure?! Do you Believe In GHOSTS?!'

It is not the way young Jedi.

It's a target audience thing. Not everyone is going to read what you write. That's ok. I tend to approach my book signings, not as a way to get a sale, but by making it my mission to make people smile at least for a moment. It makes my day a lot happier and at least I am putting out something positive into the world.

Step 7: Handing In Your Invoices & Following Up

At the end of your signings make sure you tally up the number of books you sold and do up an invoice. I try to hand them in personally to the manager at the store; it allows you to build the connection with your book seller and reminds them that you are a real person, not just a number.

Then in the mean time, when you're not there for signings, call up every month or so to see how things are tracking.

***

Advice From The Experts

When I did a launch for my previous website E-Book Revolution, I had the lovely Mark Leslie Lefebvre (previous of Kobo, now with Draft to Digital – http://markleslie.ca/) also chime in on the topic of connecting with your local community via events. Mark was a book seller for over 20 years and has been an indie author for the last ten. He has a wealth of knowledge on running local signings for his independently published books and was kind enough to let me include his blog post here in this section of the book.

I have been a writer since the age of 13. But I have also been a bookseller for more than 20 years. That might be part of the reason why I believe that there is inherent value for authors in their local bookstores and libraries.

But there's also another reasons.

The physical nature of placing books into a customer's hands.

There's almost nothing as powerful as establishing that connection between reader and author via that simple gesture.

And booksellers and librarians do that every single day in towns and cities around the world.

3.22 How To Get Your Books Into Your Local Book Store

It is far from a simple transaction. It's a layered and complex weave work of trust that is usually accompanied by supportive discussion and multi-faceted context that results in that particular book being perfect for that particular customer on that particular day.

It is truly an art. And it is personal. And rich. And deep-reaching.

As a bookseller I have put hundreds of different books into thousands of customer hands over the years. But in pretty much every single case, it wasn't just a blind automatic recommendation of a great read. It was part of an insightful understanding, not just of what a customer had already bought or read, but of their reaction to a particular author, or novel, or type of book that we had discussed, often in detail. Those complex insights helped me help them discover their next great read.

The personal relationship and the curatorial nature of the local bookseller or local librarian is key and is something that authors should be exploring in more detail. That local bookseller or librarian doesn't only know their customer's needs, but they understand that customer's passions and have often engaged in conversation regarding that customer's reaction to their previous reads.

Statistics show that customers who buy books typically make that decision by consulting trusted friends and sources. Booksellers and librarians are often among those trusted sources due to the manner by which they can cultivate deep and meaningful relationships with the people who return to them for that unique experience.

That's one reasons why Kobo, a digital eBook company, has partnered with booksellers and local retailers in countries around the globe.

Kobo sells eBooks and focuses on delivering digital books to customers. Local retailers focus on delivering appropriate books to their local communities. Combine what the two do best and you have a mutually beneficial collaboration.

And that's where you, the author, come in. The first question I'm going to ask is this: Do you already have a relationship with your local bookseller or librarian?

If you don't, ask yourself why?

When I was a bookseller, I would often go the extra mile to

help promote and hand-sell books from authors who were personable, pleasant and treated me, my staff and my store with respect. And it doesn't take much for an author to make a positive impression. But authors who came in and had nothing but negativity and a bad attitude left a terrible impression. And I rarely ever felt motivated to do anything for them.

I was particularly fond of local authors who were a part of my community and felt like a friend of my store. Whenever given the opportunity, I would go out of my way to invest in that author's success.

That's part of the karmic nature of interacting with people. Sure, when you're dealing with a website and systematic algorithms you don't need to be pleasant or polite. You push buttons and game the systems and you work the numbers in a cold and calculated fashion. But there's another way of connecting and expanding a dynamic and organic street team and bringing you a powerful ongoing ally whose main role is to move books through their system. And that's by forging relationships with real people making recommendations to real customers every single day.

The most relevant parties to consider when it comes to moving books the connection made between readers and authors. It is really all about the connection that an author makes with a reader. But they need to be brought together – and something needs to bring them together. The bookseller and the librarian bring huge value by assisting in the brokering of those relationships; they are the curators who help readers find the right authors and the right titles for them.

And getting that local bookseller or that local librarian on your side is certainly not easy; but it's definitely worth pursuing.

But you need to be aware that, just like not every reader is going to be a fan of the type of book that you write, not every bookseller or librarian is going to be in tune to your books. And that's okay. Creating relationships with local librarians or booksellers is no different than creating friendships and relationships and networking with other writers or with readers. You create those relationships by being genuine, by finding common interests, and by looking for ways in which both of you can benefit.

3.22 How To Get Your Books Into Your Local Book Store

There's no quick fix, no instant "keyword" to enter, no quick buttons to push to make this happen. It takes patience, persistence and practice – essentially, it takes a lot of hard work. However, here are a few things that authors (even digital authors) can do to help broker those relationships.

Event Ideas (with Pros and Cons)

1) Multi-author events

It's difficult for a single author to attract a lot of attention or draw an audience. Two might double the potential attention and audience, three might triple it, but even more might bring exponential value, because you're not only supporting one another and cross-pollinating each other's audiences, but you could be creating a bit of a buzz just by having a small crowd of authors. Crowds draw crowds; people tend to flock to places where there seems to be "something going on" – there's a huge curiosity element that helps; but having many authors in a single place doing readings, signings, talks provides multiple reasons for people to come check something out and also combines the many channels and outreach that each author has links in.

The total is definitely worth more than the sum of its parts. The buzz about many authors working together could also be a good hook for local media looking to share stories about events and activities taking place in the local community. One event I recently attended in Halifax, Nova Scotia was called, quite simply, Halifax Author Event and featured 30 different authors, readings, free prizes and was sponsored by a local bookstore.

2) Thematic Tie-Ins

Similar to the manner by which a multi-author event draws on more than a single author to attract attention, an event that is tied to a particular theme might be more compelling than just the fact that your new book is now out. Yes, it's exciting that you have a new book, but the average person might not be all that interested; however, they might be drawn by a theme that ties in to your book. For non-fiction, that's pretty easy, but even for fiction there are likely speaking points or thematic elements that could potentially draw interest from the public.

One example might be a novel that you have regarding Artificial Intelligence, such as Robert J. Sawyer's WWW triology (Wake, Watch, Wonder) about the internet gaining consciousness. The theme alone is interesting and focusing on the question of what might happen if the world wide web became sentient is a hook into the book. Thus, Sawyer has done talks at Universities, Libraries, Bookstores and even places like Google on the concept of this trilogy drawn from a scientific article he read that stated the concept that by some time early in the 21st century the world wide web would have as many interconnects as the human brain.

3) Goodies

A bowl of treats (such as mini candy bars, chocolates or other sweets) available for people at your author table is a nice way of allowing a discussion point that is not all about buying your book. It can act as an ice breaker and gives people a reason to approach your table. But don't just limit yourself to treats that you can eat. Consider the concept of theme above and potentially there might be something more fitting with your book that makes sense.

If you wrote a novel that features a teacher as the main character, then handing out free erasers or pencils (potentially branded with your book title/author website, etc) might make sense; if your book features a tense situation or a character under a great deal of pressure (maybe a plot point or perhaps another profession, such as air traffic controller), then maybe branded stress balls or squeeze toys might be more applicable. Goodies provide a simple ice-breaker and or conversation starter but also put the potential customer into the right frame of mind since you're offering them something.

4) Something Free (or Something to Win)

Along the lines of offering something for free, there's always the goodies or treats or promotional items mentioned above. But there could be something more substantial. One thing that I often do when

3.22 How To Get Your Books Into Your Local Book Store

doing book signings, is a printed postcard sized handout that includes one or more of my digital only books with a coupon code that allows the customer to download it for free.

For example, at a recent Halifax event I attended, I provided a coupon code allowing readers to download my digital short story "A MURDER OF SCARECROWS" for free from Kobo. There was an interesting speaking point related to the tale because it had been inspired by a town about an hour and a half outside Halifax that was known for a small army of realistic looking scarecrows that one lady was responsible for and was written shortly after I had first visited the city 6 years earlier. Most people I spoke with knew of the small town and were intrigued by it.

I have also provided cards like this to local bookstores and libraries, letting the staff who found the tale interesting (and were likely to suggest it to customers) take them to hand out to their customers however they saw fit. IE, perhaps include it with another purchase (as a free bonus) or to use to give to the people who'll likely most appreciate it – their best customers, fans of that particular genre, etc.

Along the same lines, maybe there's a single higher priced item that thematically ties in with your book (or is just plain interesting, "hot" or "cool") that you could offer as a free raffle (no obligation) that someone can win. Maybe it's an iPad, a Gift Card for the bookstore you're in, a special gift bag or a Kobo eReader.

5) Props

One thing I have done to help draw traffic, curiosity and create a bit of an ice-breaker is use thematic props associated with my book or books. Because I write horror and weird fiction, I tend to dress my table up with spooky props. Skulls, tombstones, hanging bats – and even my 6 foot skeleton sidekick, Barnaby. Barnaby helps give people an easy conversation starter; they will joke that my friend looks like he needs something to eat, they'll ask if he'd been sitting around for a long time or they'll just ask what he's about. That helps break the ice. Another value that Barnaby brings is he helps people understand quite quickly, often from afar, the style or genre of the book(s) being presented, allowing them to decide it's not for them and stay away, or perhaps drawing them in because it's exactly the type of book they like.

Barnaby now has his own Pinterest Page along with other social media and has become such a popular part of my author branding that when I am invited to attend particular events, a common question is whether or not I am planning on bringing Barnaby along (including when I was invited to be a guest of a local television morning program).

Sometimes, any of these ideas for events will give the local bookseller or librarian additional information to share to customers about you. Perhaps you were one of the authors at a memorial multi-author event, or your props led to an amusing anecdote that people enjoy listening to – whatever it might be, that just might become additional fuel that will last far beyond the one day that you were physically there.

As mentioned, cultivating a relationship and coming up with ideas for events at local bookstores and libraries takes a significant amount of work. But when you can forge that relationship, when you can have one or a dozen booksellers or librarians out there who believe in your writing, believe in your books and are eager to recommend your books to the right customers, you have mined an incredibly valuable renewable resource.

3.23 How To Structure An Author Talk

The days of writing in a garret and passing on your words, never to look a reader in face, are done. Overdone. Well done. They may quite possibly be charred beyond recognition and masquerading as a black hole. Or charcoal. In fact if you're a writer, and you get a stocking full of charcoal for Christmas, then they are your old writing garret days. Sorry.

Because the fact of the matter is, if you want to make it as a writer, and I mean *really* make it as a writer, you have to be able to learn to connect with people. And not just for the warm and fuzzies and building your tribe. Yes that's part of it, but that's not the main reason. We can't eat warm and fuzzies. We can't pay our bills with good will and pats on the back. Sure it builds the confidence but my preference is not to get to the point where I have to eat paper to feel full.

Many authors don't realise this, but the money an author gets for speaking for an hour, or giving a workshop, or appearing at a festival, is what sustains full time writers in their careers. Even well know and successful indie authors like Joanna Penn admit that it is their speaking gigs that keep them afloat, not the sales of their books. It's here that we come to a problem. Because if you suck at public speaking, those gigs are going to be as likely as flying dolphins with jetpacks. Even if you are a confident and mildly amusing person, if you don't know how to do it right, if you don't know how to flip those switches like a Japanese game show host, if you don't know how to reach all types of people, chances are you are only going to connect with 5% of the room. If you don't know which 5%, they'll be the ones that *don't* have that glazed look in their eyes that says they've been watching reality TV for too long.

I've been to a lot of bad presentations; I think I've killed at least half a dozen authors in slasher daydreams in my head, and I decided that that wasn't going to be me. So last year I decided to take a three day course in the art of public speaking from a wonderful woman called Carren Smith (*https://carrensmith.com/*). Carren is an amazing public speaker, a beautiful woman, and an even more amazing human being, coming all the way from surviving the Bali bombings to where she is today. Before that weekend I would never have known the art of opening your mouth and making sounds could be so calculated but at the same time, so simple. I'd been doing it all wrong. If you get the chance to attend one of her seminars, you could not go wrong.

In this section I'm going to break down the mechanics of a short presentation for you, so that you too can learn how to structure and land those speaking gigs that allow you to start carving a writing career that *makes* you money, rather than sucks you dry. In fact this structure works so well I've started using it for my blog post writing as well. Your talk can be on a topic or theme from your book, it can be about how you write and teaching others to be creative. It doesn't matter *what* you want to get across, we'll have you ripping up the stage like a bad ass entertainer in no time. So let's get too it!

There Are Four Different Types Of People

No I don't mean fat, thin, tall and dwarf. I don't even mean greedy, kind, mathematical or good a selfies. It would be more accurate to say there are four different types of learning styles. If you cover all of the different ways people learn in one presentation, then you are more likely to reach *more* people and suddenly you're seeing your 5% attention rate jump up to over 50%. The four learning styles include:

- **The Why People:** They want to know why they should give a damn, about you, about your book, about the life, the universe and everything. They are the ones with the short attention spans, they are the ones you have to grab first before they start throwing spit balls at the other attendees.
- **The What People:** What is this thing/object/topic that you insist on talking about. If they know *what* it is, they can classify it and are happy to move on.
- **The How People:** How do you do that thing you do? No really, *exactly* how is it done, how can they replicate that, how can they learn from all your time wasting mistakes and make a clear path for themselves. These types of learners

need a structure, a yellow brick road they can skip down with a childhood hallucination on each arm.
- **The Do It People:** Cool, cool, sweet, sweet, now can I touch it please??? Pretty please?? I need to play with it. I know that came out wrong, but you really need to let me do the thing you were talking about now, please. These people aren't happy unless they are putting something into action. They'll listen to all the talk, just as long as you give them something to tinker with at the end.

You need to address each one of these learning styles in each section of your presentation. 'How many sections should I have?' you ask. Well, there should be Four:

The Introduction

This is where you hook in all your learning styles in one fowl swoop. This section should never be over a minute and a half, if you were doing a three minute video you should have this baby wrapped up in thirty seconds. It needs to be sharp, punchy, and have all the kick of a black belt. It should contain:

- **Topic:** What is this whole presentation about? People need to know WHY they should even bother listening to a minute more. Where is this hour going? When can I have tea?
- **Audience:** Who will this presentation help? Who is it aimed at? The worst thing you can do as a presenter to your audience is to let them get to the end of the presentation none the wiser about whether or not you were speaking to them or the person behind them. They need to know, you know, who you're talking to. Basically this is the, 'Are we on the same page' test.
- **What Keeps Them Up In The Night:** What is it that has them jerking awake in their bed and nibbling at their fingers until they are bloody raw stumps? This needs to be emotional and trigger their imagination and inner feelings. Dramatic faces are recommended.
- **What in it for them:** By the end of this presentation, after

they've put off their smoko, and rested their Facebook thumbs, and decided to spend their day inside with someone they've never met but kind of looks like their smelly neighbour, what will they have? What knowledge or process or fun gift pack will they come away with?

About Me

The audience needs to know they are in safe hands and that you are in fact a human who feels human-type things and has had the same or worse shit happened to you. This is the 'Are you a robot/crackpot/narcissist/Animal/mineral/vegetable' test. If they feel like they know you, if they feel they have a connection with you, then they are more likely to trust that you will lead them into the land of good writing and entertainment. This section has a more simple structure than the first. You have to take them on a rollercoaster ride: start high, go low, end high. You're low doesn't have to be death or destruction, depression or homicidal tendencies; it just has to be a point in your journey where you could have give up but didn't and broke through. Normally it is the tipping point, the pivotal moment where you said enough is enough. It is *always* important that you end on a high note. Any presentation that starts off all doom and gloom with no way out is going to get you lynched mobbed at the end, not a standing ovation. This section should take up about 20% of your presentation time.

How To

Alright, I'm interested in what you're saying, and now I trust you, please guide me oh messiah to the promised land! Fifty percent of your presentation happens right here, right now. Depending on the length of your presentation you should have either 3, 5 or 7 steps. Odd numbers have been psychologically proven to stick in the head more (says Dr Craven…) and as a result odd numbers are used often in marketing. But add too many steps at your peril; you don't want your audience getting overwhelmed and felling like you're trying to ram a university degree down their gullet. You need to count down each step starting with the least important and working up to the most insightful-piece-of-information-that-has-ever-graced-the-world-stage! Aherm.

Anyway, as you may have guessed, this section isn't just for the how-to learners, you need to hook in each learning type into each of your steps. So for each step you're going to have to explain *why* it is important, define *what* it is you are referring to and describe it, and tell people *how* they can achieve their own Step 3/2/1 and end with one thing they can do when they get home (or do in the workshop).

The Close

Here is where you get people to take action, this is for the DO IT types, the active people who have probably been shifting in their seats like a child who really needs to pee but has been told to stay put. Whether you direct them to do a task, buy your book, or do further learning with you, this is the action point and it should involve at least a bit of body movement and have some sort of deadline attached. Nothing gets the action going like a deadline. Say you wanted to send them to a webinar. Your call to action should follow a similar pattern to the other sections, go into *why* the watchers should attend the webinar, *what* the webinar was for, *how* they could book into it and the urge them to bloody well *do it*!

The Number One Rule Of Presenting??

No PowerPoint. PowerPoint is like a growth, a crutch, a carpet pulled out from underneath your feet. When you use PowerPoint treat it like a photo frame, no words, pictures only. PowerPoint puts people into a passive state of learning. You want them in an *active* state, where they are more awake and retain more of the word-sounds that come out of your mouth. If you must, give them a sheet at the end that has all the information they need.

3.24 Crowd Funding 101 For Writers

I believe I've spoken about the cocktail fantasy before. You get your first book published (either traditionally or self) and suddenly you find yourself daydreaming about monthly holidays in the Caribbean and writing from your million dollar villa in Spain. And in your excitement you quit your run-of-the-mill job and suddenly find yourself up Poor River without your Pay Check Paddle. If only you had the money/ the apartment to get the second book off the ground that your fans so *desperately* want. Or at least you hope they want.

If only you had the money to get your next project off the ground. Most of us would suck it up by getting a cheaper, less striking cover; by agreeing to the publisher's demands to be less 'truthful' in our next book if we want it to see the light of day (and snag the much needed advance); or by relying on a family friend who's good at English to edit your novel rather than the professional you know you need.

However, if we stopped to have a look around we would realise the landscape is changing and it's all due to the readers raising millions of dollars around the world for writing projects.

What if I told you, you could get paid for writing, or to make your first print run? Especially if you are an already established author? What if you could make a couple of grand to get you started? What if you could make a million? Three million? All before finishing your final novel/magazine/comic/video game (yes, I would argue they are ALL forms of storytelling).

It's incredible to think you could get paid before writing a single word, but if you are willing to do the work, then perhaps crowd funding is for you. Crowd funding allows artists to raise funds for their project by asking everyday people to pledge money toward it. It's not like traditional fundraising where you wear a Santa suit, put out a bucket to collect change and you hand it into the organisation (or I suppose yourself in this case). Crowd funding sets a goal and time limit, if you don't get enough people to pledge in the allotted time, everyone keeps their money, and the project goes back to square one. If the artist makes the funding goal only then will the project go ahead, and everyone who pledged gets some sort of reward for their wonderful help.

In the same way e-books have allowed authors to release books in niches too specialised for a publisher to bother with, crowd funding democratises the process of story creation by allowing readers/consumers to support stories they want to see developed and give the artists the freedom to experiment, take risks and design without anyone else compromising their vision or telling them they are too boring for the This-Is-Your-Life documentary they're planning. It's a kind of creative luxury that most major, established publishers and magazines simply cannot afford to give their authors. Thus we are fed the same staple without a chocolate brownie to delight our entertainment tastebuds. We are being fed the content without getting the 'rewards' we truly crave and that crowd funding has brought into being.

Crowdfunding is the type of platform that I used to fund a photography project to Cambodia and what first time author Alain Guillemain used to edit, print, and launch his book *Customer Delight*. Yes you heard me right, Alain was a first time author, not an established author, and in 40 days he raised $7600. I was so intrigued I interviewed him about how he did it.

Lucky for you, I recorded it! You can find the full podcast interview (#4 in the list) here: *http://www.cravenstories.com/books-and-more/podcasts/ebrpodcast/*

***

It's platforms like Kickstarter (US) and Pozible (Australia) that are allowing writers to raise insane amounts of money for their projects. Crowd funding is being used at all skill levels, from first time authors like Alain, to experienced veterans Kristine Rusch and Dean Smith – awarded World Fantasy Awards for their work with the Pulphouse Publishing Company – who raised $14,000, double their original $6,000 goal, to begin a bi-monthly fiction anthology, Fiction River.

Yet, the real leaders in this crowd funding phenomenon are not the prose writers, but the authors of graphic novels and games. Rich Burlew, author of a self-published web comic The Order of the Stick,

3.24 Crowd Funding 101 For Writers

overshot his original $57,000 goal to raise $1.2 million to bring his stories back into print; while Tim Schafer and his gaming company Double Fine, overshot their original $400,000 goal to make $3 million to develop their next game, a classic point and click adventure. If you aggregate all the successful Kickstarter graphic novel projects together, it results in the Kickstarter conglomerate being the second highest grossing comic/graphic novel publisher ($1.99m) behind Marvel ($2.7m) according to Publisher's Weekly back when the Kickstarter reached it's goal in 2012.

Rich Burlew raised more then $1mill to publish The Order of the Stick.

In a previous chapter I talked about comic book companies being the geniuses of our industry. Not only do they mass produce paperback copies of their stories but they have television shows, movies, yearly conventions in major cities around the world, figurines of major AND minor characters, and heck, they even have lunch boxes. Depending on when they're made, how rare they are, and whether or not the buyer has resisted temptation and left the item in its original packaging, the figurine of a villain's hairless cat could go for several hundred dollars when first sold and several thousand dollars years later. Graphic novel author and artist Dan McGuiness, creator of Pilot and Huxley, explains this reader-creator model best:

"Comic book stores haven't suffered the way book stores have with the digital revolution for one simple reason: Readers of comics are collectors. Comic series are built on rarity, on the knowledge that their fans want a special connection to their story and different ways to be a part of it, whether it is the rare first editions or figurines. I know several comic themed magazines that have gone out of print because comic news has gone digital, but because of that relationship between brand-series-reader, the comic store thrives. That's probably why the e-book aspect of comics has started off so small, giving limited options to go indie for new artists."

Graphic novelists (aka comic artists) have the same publishing restrictions as prose authors; there are only a handful of comic publishers in the world in comparison to book publishers, making the competition to be picked up by a house fierce. Most artists resort to free webcomics to be discovered, or to gain a following of their own so they can sell independent print runs. Many, including Rich Burlew several years ago, have to give up the printing of their comics due to cost. It seems crowd funding is one of the best avenues of indie publishing for a graphic novelist. This is because the crowd funding model fits the mindset of comic readers so well; it is aimed at this collector mentality, aimed at the school of thought that recognises a reader just wants to be involved.

One of Double Fine's many rewards for its game development project on Kickstarter.

It is the reason why these two projects have made millions while writing projects have only made thousands. "People want to own what they love," explained Burlew, "So rather than selling access to the content, sell the permanent incarnation of it." Exploiting that advice is something these projects have done brilliantly. Burlew offered a variety of awards including: an Order of the Stick fridge magnet, digital PDF of the full story, books, prints, autographs, an original crayon drawing, and if you pledged $5,000 your original Dungeons & Dragons character would receive a walk-on cameo in the webcomic. All options were sold out.

The Double Fine gang went a step further in their project and promised an exclusive experience (see video below). The backers would get to see the game development unfold in real time. The whole creative progress was to be documented and released in monthly video updates exclusively to Kickstarter backers. A private online community would be set up for the backers to discuss the project with the developers, giving their feelings about the game's direction and even being able to vote on some decisions. All of this was provided for a base pledge of $15. Anyone who pledged at levels higher than that got access to extras such as the full documentary in HD with bonus footage, unique posters, original concept art, or even a mini portrait done by the game's artists. In essence they tailored their project for the readers, who were interested in the workings of the game industry. This concept saw them funded 750%.

These types of projects are what the crowd funding platforms encourage, pushing the idea of creative rewards hard. It's little wonder that indie publisher Richard Nash talks most eloquently on writers needing to expand their scope from the novel to further interactive opportunities like workshops, Q&A sessions, memorabilia, exclusive dinner parties, their own board game or selection of swim wear (ok, so those last two were my ideas, but you never know…). Authors need to embrace the realisation that the comic and gaming industry have exploited for decades, people want to own what they love, but they also want something special and something that connects them to the creator. If authors and even independent publishers wish to float cutting edge ventures and command the kind of money crowd funded by these projects, they need to shift their mindset. A story is more than just a book; it's the chance to connect with an audience in an infinite web of possibility.

Crowd funding is the ultimate test of audience interest and connection. A bad idea or project will not float. The power of this method is now being acknowledged by government organisations like ScreenWest, who are matching the funding of crowd funded films in Western Australia three to one. So for every dollar raised, ScreenWest gives the successful project three bucks.

It's easy to forget with the large numbers flying around that funding a successful project is not easy; it is a professional, artistic endeavour which requires a lot of planning, marketing, audience connection, and the ability to produce what you promised. Needless to say, proposing the purchase of sixty watermelons, a trebuchet and a giant canvas for experimental artwork is probably not going to see you raise five cents let alone three million dollars. Almost 50% of projects *do not* get funded in the allotted time. So you need to have a good, creative plan in place, a fan base and dedication. It's also best to accept the cold hard fact that comic artists and gamers are geniuses, whose creative brain prose writers need to clone.

3.25 Interactive Storytelling: Real-Life Choose Your Adventures

Who hasn't read a story where they wished they could actually be inside it? See the landscape for yourself, see how the light falls, the air smells, the noise overwhelms you and see exactly how tall that building was that Spiderman just scaled. It's not that we don't trust the author and their powers of description, it's just we want to be there not just read about it, and ultimately we want to tell our own stories of what it was like.

I'd been obsessing over the idea for a while in 2012. Just how could you *do* it? Then I heard Simon Groth from If:Book Australia (The Institute for the Future of the Book run for several years out of the QLD Writers Centre) talk about an app that was being developed. The story came in locked segments and if you wanted to unlock the next bit of the story you had to be in the place where it happens. Say for example the next bit of the story happened in a train station. You don't have to be at the exact station in the story (e.g. Grand Central Train Station in NYC) to unlock the next bit, you just have to be in *a* train station somewhere. As long as there's a Thomas the Tank Engine near you, you can merrily read away. It was reasoned that being in the right atmosphere made it feel more real.

While that is one of the most awesome ideas to come out of this digital era, I did not have a million dollars to spend developing an app. In fact, if I had a million dollars I would assuredly have quit my job and be writing in a villa in Italy right about now. So my next question was, how could a *normal* person do this? Then I realised I wouldn't just want to move from place to place following the story of another character. I wanted to call the shots. I wanted the options. I wanted to choose my own adventure!

After six months of pondering Adelaide: Choose Your Own Adventure was born. The project involved the world's first (yes, I Googled it) Choose Your Own Adventure event. Rather than reading the CYOA in printed book form, the project placed QR codes around

Adelaide city that you could scan with your smart phone. The code links you to the next part of the adventure where you can choose from several options to continue the story. Each new part of the story took place in the location of the QR code, showcasing Adelaide city landmarks in a whole new way. The adventure started from a single point in Rundle Mall during Adelaide Writers' Week then branched off into three separate stories by three authors: A comic alien invasion of Victoria Square (Emily Craven), a Sherlock Holmes style mystery in the East End (Henry Nicholls), and a dark thriller where the city facades came to life around you (Ben Mylius). The posters looked like this:

What truly fascinates me about the variety of projects in this new age is exactly how they came about. What ideas bubbled up to make it all come together? Because the more you know about other people's creative process, the more ammunition you have to throw off your own shackles of impossibility and create something really interesting. Here was my thought process for ACYOA:

3.25 Interactive Storytelling: Real-Life Choose Your Adventures

PROCESS 1

Ignite curiosity. How cool would it be if I was actually *in* the story? No, not like a picture book. Actually there!

PROCESS 2

Attend an e-book seminar, do your own research, attend a three day internet marketing seminar that has nothing to do with publishing or books... at all, start a blog about ebooks/author marketing/connecting writers-readers.

After If:Book's wonderful e-book publishing seminars with Mark Coker from Smashwords, I set about learning all I could about everything digital. My father was setting up an online training business for hairdressers at the time and dragged me along to an internet marketing seminar. It was one of the most fascinating seminars of my life. Focusing mainly on making money online, the tools that were being used by the big players were all common sense techniques that could be applied to selling anything, including books. As a way to sort it out in my head I started a blog, running it over 31 days with a new concept from the seminars being explored each day. I also signed up for the mailing lists of several experts and companies.

PROCESS 3

Read the emails you signed up for.

It was the email on how to use QR codes that set me off. A QR code is a 2D barcode, generally square in shape that can be read by barcode apps on smart phones. Before falling out of fashion they were used on various promotional posters, on the side of Pepsi cans, or on ads in the subway. When someone scanned the barcode with their phone it took them to a website. They are mainly used by marketers to promote a company. But they had so much more potential. For this project, the website contains the stories for each step of the Adventure. At the time, in my mind the best thing about a QR code was the story could be a long as you liked, because it didn't have to be printed and if you were really daring you could merge it with YouTube videos and music which you would never have been able to print on a mere poster.

PROCESS 4

Mix up your reading.

I remember choose your own adventures from when I was a kid. I'm pretty sure I still have a Star Wars one tucked away in a closet somewhere. It was a little too serious for my taste and I always died within the hour. How much fun can you have when you are dead? As fate would have it I'd switch from novels to short story reading for a while in the summer of 2012. One of my favourite authors, Garth Nix had a collection called Across the Wall, and lo and behold there was a choose your own adventure called Down to the Scum Quarter. It is hilarious, and if you die, you die with tears of laughter.

As fate would have it, I was tackling QR codes at the same time. And the rest, as they would say, is a dark alien invasion Sherlock Holmes thriller.

***

When people harp on to me about the smell of a book I have to say I'm fairly dismissive. I mean come on, when was the last time you saw someone on the train taking a good whiff of a book as if it were a bunch of pungent flowers? I pick a book up because of the content, the author and yes, the pretty front covers. I pick it up because I know it's going to take me somewhere new, and if I'm lucky, the author is going to have a little play with their words and format. Digital will work just as well for this as any musty book that makes you cough if you sniff a little too hard.

The digital era is allowing us to do so many things with the written word, creating new forms and genres. It also has the capacity to bring an old art form back from the literary dead, such as the Choose Your Adventure. Let's face it, we're never going to see the Choose Your Own Adventure (CYOA) back in print, but digital has allowed us to resurrect this childhood memory, and the writing skills that were lost with it.

3.25 Interactive Storytelling: Real-Life Choose Your Adventures

It may surprise you to hear me talk about writing skills and CYOA in the same sentence. Perhaps you believe that CYOA is just a form of fan fiction gone mad, requiring no writing skills other than to be able to string words together. However, the truth of the matter is, writing a CYOA? Not as easy as you might think. Creating multiple endings is exceptionally difficult, for as writers we rarely contemplate two or three endings let alone the eight endings I wrote in my Victoria Square Invasion for the city of Adelaide, Australia. And you just can't make the endings subtly different (what I would term a creative cop-out), because why would your reader bother to choose their own adventure when their choice is no choice at all? Also, when you're doing a physical CYOA you can't just make up the location details and props as you would a book; they have to have some basis in the reality that surrounds the participant/reader.

In the end, when our creativity failed us, the locations of Adelaide kick started our brains. We had to start digging into quirks of Adelaide, the locations where interesting features had gone unnoticed by the normal pedestrians. What was it about the location that could set the scene? What might happen in the story to bring the reader's attention to this feature? Did the location itself have an atmosphere that could be played on? Was it close enough to the previous location to stop our readers wearing holes in their shoes? We also had to consider whether we were going to make the endings wins, loses, or partial wins. Making a person walk from one side of the city to the other and then having them die, may see your project unattended the next year. Like in all writing, reader satisfaction is key.

In a way, these restrictions were a godsend, because they got us out of our writing comfort zone – out of our writing fat pants and into the sweat pants. Elements that I would never have added normally were incorporated because of their unusualness rather than being dismissed as too unwieldy. In the long run it made the whole experience and the adventure itself, unexpected and more interesting, because these things were *there,* the reader could *see* them, and it made the science fiction component more believable.

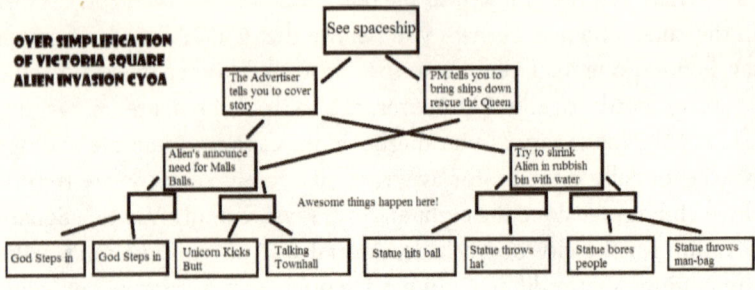

Now, I can apply this skill to my normal novel writing. So rather than ending with the easiest and obvious path, I can contemplate ends that are vastly different and, in many ways, more satisfying. John Cleese, in his wonderful address on creativity for SMI would say that I am allowing myself to 'play' more. To root around for the right answer rather than the noticeable one. To give a contemporary example, I would point to Suzanne Collins who wrote the Hunger Games. Though I didn't like the way the series ended, I knew it was the *right* ending. It's an invaluable skill to have and one I would never have thought I needed before I embarked on this project. I've always been a 'this idea WILL work' kind of girl. Now I have a story that involves shrinking aliens in water; four statues and a Post Office façade of a unicorn, coming to life and battling a spaceship; and a choir of rather dirty angels bringing down God's wrath on alien invaders. Adelaide has never seen so much drama.

This project across the past couple of years has morphed into a locative literature app called Story City (which I tell people is Pokemon Go, but with stories). Some how it has become a tech start-up that empowers writers, and creators across the world and has won several awards. We want to redefine stories so that they are a part of rather than something you're told. Each story is accessed via our iOs and Android app, which only unlocks a story when you're standing in the right place. Whether they are real-life choose-your-adventures, puzzle trails, indigenous myths or historical tales, our stories are created by local storytellers and artists who know their city best. Part of our ethos

3.25 Interactive Storytelling: Real-Life Choose Your Adventures

is we don't bring people into a town to create the stories, we provide professional development to the local community to create interactive stories so THEY can create the tales for their cities and towns. Now the Story City app has 40+ stories in half a dozen states in Australia, international walking tours, and has trained over 250+ creatives in interactive storytelling.

With Story City the idea is to boost and support our local creative economies in a way that provides opportunities for our local talent and a way for them to have a sustainable career without needing to move away to do this. We wish to help create opportunities for cultural rejuvenation and tourism in cities by providing the training and tools for communities to create this digital storytelling and gaming layer to engage locals and tourists alike with public spaces. We in essence add a digital cultural layer over the real world, and inviting people to see their everyday spaces through a different lens. This is why my focus has switched from writing novels to being the CEO of Story City, there is an opportunity here to impact the world with thousands of stories rather than the dozen or two that I could ever manage to write. While I'll always be writing my own novels, I know that this collaborative work makes my heart sing just as much, and my publishing and editing skills have galloped ahead so much further than I could have ever imagined as a result.

I'm super keen for YOU to be our next adventure writer, so if you want your city to become its own story, get in touch www.storycity.com.au.

Umbrella 3: Important Chapter Links

Podcasts

- Crowd funding podcast: *http://www.cravenstories.com/books-and-more/podcasts/ebrpodcast/*

Article Writing

- Article writing website: *www.ezinearticles.com*
- Article sharing website: *www.digg.com*
- Article sharing website: *www.alltop.com*

Communities

- Google Groups: *http://groups.google.com*
- Yahoo! Groups: *http://groups.yahoo.com*
- Kindleboards: *www.kindleboards.com*
- Reading community: *www.goodreads.com*

Paid Advertising

- Google Adwords: *www.adwords.google.com*
- Amazon KDP Advertising: *https://advertising.amazon.com/kdp-authors*
- Facebook advertising: *https://www.facebook.com/business/ads*

Press Releases

- Press release outlet: *www.prweb.com*
- Australian press release outlet: *www.medianet.com.au*

Social Media

- Twitter: *http://twitter.com*
- Searching Twitter: *http://search.twitter.com*
- Facebook: *www.facebook.com*
- Create a Facebook page: *www.facebook.com/pages*

- YouTube: *www.YouTube.com*

Website Creation

- Squarespace: *www.squarespace.com*
- WordPress: *www.wordpress.com*
- Hosting/domains: *www.bluehost.com*
- Hosting/domains: *www.hostgator.com*
- Hosting/domains: *www.godaddy.com*
- Domains: *www.namecheap.com*
- Website development freelancer websites: *www.upwork.com*
- Website development freelancer websites: *www.freelancer.com*
- Blogging platform: *www.wordpress.com*

Other links

- Bookfunnel: *https://bookfunnel.com/*
- Pay with a tweet: *www.paywithatweet.com*
- URL shortener: *http://bit.ly* or *http://tiny.cc*
- Article website for Newsletter Content: *www.ezinearticles.com*
- Email List Auto-responder: *http://tiny.cc/EBRgetresponse* (affiliate)
- Hidden Gems reviewers: *https://www.hiddengemsbooks.com/arc-program/*
- Netgalley reviewers: *https://www.netgalley.com/*
- Luke Gracias's Post on using Netgalley for his book: *https://fiveplustwoblog.wordpress.com/2017/02/03/netgalley-for-a-self-published-debut-author/*
- How NOT to handle feedback: *http://booksandpals.blogspot.com/2011/03/greek-seaman-jacqueline-howett.html*
- Subscription websites: *http://www.astorybeforebed.com*
- Affiliate promotion: *www.clickbank.com*
- Smashwords affiliate promotion: *www.smashwords.com/about/affiliate*
- E-book Revolution Interview Series: *http://www.cravenstories.com/books-and-more/courses/interview-series/*

Umbrella 3: Important Chapter Links

- Skype to conduct interviews: *www.skype.com*
- Free editing softwear: *http://audacity.sourceforge.net*
- Video editing softwear: *www.techsmith.com/camtasia*
- Webinars: *www.gotowebinar.com*
- Google Hangouts: *https://hangouts.google.com/*
- Video promotion example – Go the F*** To Sleep: *https://vimeo.com/26403238* and *https://youtu.be/Cb0t9TUNLpg*
- Handmade Tolkien book: *http://tiny.cc/TolkienArt*
- E-book launch for Isobelle Carmody's Greylands: *http://tiny.cc/greylands* & *http://tiny.cc/greylands2*
- Mark Leslie Lefebvre (From D2D): *http://markleslie.ca/*
- Carren Smith (Public Speaking Coach & Motivational Speaker): *https://carrensmith.com/*
- Story City choose your adventure project: *www.facebook.com/StoryCityAdventures*
- Emily's About Me/My Story page: *http://www.cravenstories.com/my-story/*
- Emily's Create a Newsletter Course: *http://www.cravenstories.com/books-and-more/courses/*

The Revolution: An Ending

The enemy of a writer is the veil of obscurity that lets our words slowly collect dust as they are transferred from one slush pile to another. Though initial doubts about the digitisation of our industry can cloud our thoughts, one thing is clear: publishers do not know the minds of every reader, and they cannot afford to put all of us in print. E-books have so many benefits, and the internet is a medium through which we can connect on a whole new level; not to mention the opportunity to make more money than a traditionally published author ever did.

Yet by far the most appealing aspect? As an author, you have the ultimate creative control.

Want to stay up-to-date with my industry musings and creative ideas?

Sign Up for my Author Newsletter at

http://www.cravenstories.com/books-and-more/freebies/

If you found this book helpful, please review it!

Reviews are the life blood of authors, without them our books flounder. So please, if you found this book helpful please review it on Goodreads, your favourite e-book store, or even your own blog! Many thanks.

Goodreads: http://bit.ly/GREbookRevolution

Have You Found This Book Invaluable?

Then Expand Your Skills With Me!

Work With Me

I work with the dreamers and the storytellers, the rule breakers and the quirky. I work with a new breed of creatives. Writers who feel it to their bones that collaboration is the best way to take their story from just another tale, to changing lives. If your aim isn't to connect with your whole heart to your audience, if you're aim isn't to refine the story of your life or next novel so that it sparkles like the flaming arc of a shooting star, then we're not a fit.

I show daring creatives how to draw the curious down the rabbit hole with stories, how to use their tales to spark connection, understanding, and create belonging with a wonderland of their making.

Yes, all of you are capable of creating a wonderland with your stories! But first you need perspective.

So, if you shudder every time you have to explain the stories you write…

If there is a disconnect between your head and the page, your personality and your words…

If you know that somehow, something is missing in your story because people aren't seeing your magic….

Or if your stories are like the Swedish Chef's spaghetti on Sesame Street (a big hot mess!), then let's untangle that bundle of crazy together and make those stories sing like you've switched bodies with a 20-year-old Julie Andrews.

Because when you work with me, I work *with* you and your story, not on you. Dive down the rabbit hole with me: http://www.cravenstories.com/work-with-me/team-writer/

Training For Writers & Independent Publishers

Get the information and knowledge you need (without the time suck) with these Super Writer Multimedia Courses. These courses cover, step-by-step, what you need to know about self-publishing, e-books, building an online platform and connecting with readers, and are designed to inform while at the same time *not* break the bank balance.

Creating an Email List For Writers

Start building your greatest biz asset! That invaluable email list of readers.

Ask any successful independent author; they will tell you to build your email list months before you even think about releasing a new work. This course shows you how to do it in a way that feels right to you! Find out more about the course here:

http://www.cravenstories.com/books-and-more/courses/email-list-for-writers/

E-book Revolution Interview Series

Not sure if this whole go-it-alone, independent author thing is right for you?

This course gives you the low-down on the publishing landscape, ebook and print creation, as well as unique marketing ideas with over nine audio interviews, transcripts and much more! Find out more here:

http://www.cravenstories.com/books-and-more/courses/interview-series/

More Books By Emily

You can purchase any of these from your favourite ebook store. Just follow the links from my website:

www.cravenstories.com/books-and-more/books

Original Fantasy: The Practical Guide To Writing Genre

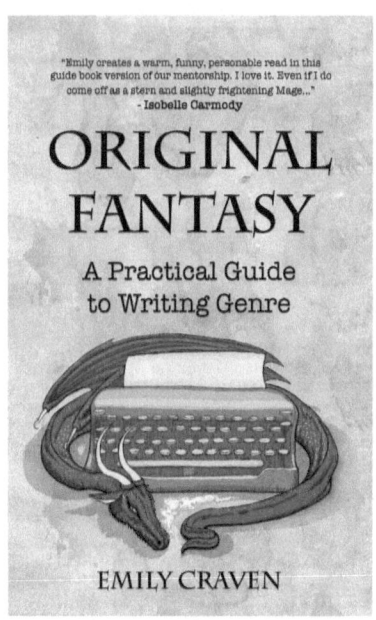

"Emily creates a warm, funny, personable read in this guide book version of our mentorship. I love it. Even if I do come off as a stern and slightly frightening Mage..." ~ **Isobelle Carmody**

So you've finished your novel and it is Frankenstein's monster. Don't Panic!

Whether you write fantasy, science-fiction, crime, thriller, YA, or chick-lit, this humorous and helpful guide will hone your skills and lead you through the quagmire of writing original fiction.

Emily Craven details the lessons she learnt during her twelve-month mentorship with award-winning author Isobelle Carmody. Emily has included dozens of examples of her original passages, along with Isobelle's insightful comments. An award-winning publisher and author in her own right, Emily simplifies and demystifies the complexities of writing and editing your novel.

Madeline Cain: The Adventure Begins

Addicted to Facebook? Love YA? Ease into The Grand Adventures of Madeline Cain series with the free novella, *Madeline Cain: The Adventure Begins.*

Life after high-school is looming and Madeline Cain is freaking out. Everyone has an opinion on what she should do with her life but her. What sane person decides the rest of their life at 17?

As Maddie resigns herself to six months of decision-making hell she meets Claire; an exchange student from Ireland with a wicked sense of humour and an aversion to technology. Claire convinces Maddie to join her '365 Days of Fun' project and suddenly Maddie finds herself giving fake tarot readings at the beach, dressing up as a superhero to stop petty crime, and hijacking a cult from its creator. But when Madeline gets caught 'playing games' rather than taking her future 'seriously', reality comes crashing down. Will Maddie find the career of her dreams? Or is she doomed to spend her life adventure-less?

Written as though you are reading Madeline's Facebook page, *Madeline Cain: The Adventure Begins* is a contemporary comedy that will leave you in stitches…

Jake's Page: A Short Story & Play

Jake Black is almost your average teen, fresh from the high-school farm he jumps at the chance to move from Hobart to Adelaide, with the vague plan of taking engineering at the University. If only his mother wouldn't call him every day to check on his diabetes. Or use his sister's Facebook account to spy on his activities. However college is more adventurous than Jake's parents bargained for. From his first 'ponding' in the courtyard fountain to forced o-week activities such as Goon of Fortune, and being dropped 30km from the city wearing nothing but a g-string and body paint, Jake plunges into college life with the enthusiasm of one finally release from the parental blanket.

Ignoring the inevitable medical impacts of his new life style Jake struggles managing his mother and his new freedom with mixed results on the public Facebook platform. Jake's Page is a comic yet heart-breaking tale of a young type 1 diabetic, and his Facebook Page, told through status updates, Facebook notes, and Facebook personal messages. This e-book presents the tale as both an short story, and a play.

About The Author

Chocolate. Karaoke. Star Trek. Travel. Books. Puppies. Shaking what your Mama gave you. All of these are some of my favourite things. But when I meet someone, I want to know who they are, not what they like. I want to know what's their story? Why do they get up every morning? Other than, like, needing to have a pee.

Aherm, moving on.

For me, what rocks my world is showing daring creatives how to draw the curious down the rabbit hole with stories, how to use their tales to spark connection, understanding, and create belonging with a wonderland of their making.

Stories entered my DNA as a kid. They were what saved me from lonely lunch times with no friends when my family moved states and I was shoved into a new school mid-year, mid-puberty, mid-awkward-phase. They allowed me to escape to another world of adventure, of struggle (that wasn't mine), of empathy, perspective, and heroes who strived against the bullies, and again and again, picked themselves. Stories showed me how to adapt, to care, to trust myself. They understood me on a level I barely understood myself. I was such a voracious reader I started writing my own books when I was 12 because my favourite authors just couldn't keep up.

Stories were how I survived boredom. Boredom was how I ended up a Star Trek nerd. Every afternoon when I got home from school, my mother commandeered the TV to fuel her Star Trek addiction. The choice was be bored or be obsessed. You could say I was brain-washed a Trekkie and I have no regrets!

That's the only reason I can think of for how I ended up choosing to study Astrophysics. Two years in and something happened that I never in a million years expected. I hated it. I had no idea what else I would even do if I quit. I was good at it, sure, but every six months I would have a mini-break-down in my bedroom, the words of high-school teachers and parents going around and round my head – 'you're too smart for art.' If present me could time travel, I'd go back and slap them all up-side the head, with a loud, 'fuck that noise' for good measure.

How many times have you been told you 'should'? You should do this, you should do that, even though you know that box doesn't fit you?

What I didn't realise at the time was the reason I was so drawn to Star Trek wasn't the science, it was the adventure. A soap opera in space; people working together solving problems, falling in love, and shooting phasers! This was the root of my unhappiness; I was suppressing the biggest part of myself. I didn't want knowledge for the sake of knowledge, I want to create things that connected people.

And the way that excited me, that lit a fire in my belly to create that connection, was by creating and sharing stories. Fictional preferably, with a hint of magic, a dash of quirky, and a sneaky side of truth.

I wish I could tell you that when I set my sights on career as storyteller, I shook off that 'should' energy. I did not. While I devoured dozens of courses on writing, publishing, marketing, editing and eBooks, and learnt one of the most important lessons of my life – that what you create alone will never be as good as what you'll create together with the feedback of professionals who aren't you and see your blind spots – I was still doing all the things you should. You should send your novels to traditional publishers, you should write short stories to get a name for yourself, you should have a 'very' professional website where you're 'very serious' and therefore 'competent', as confirmed by your head shot which makes you look like you have sat on a cactus.

I waited a really long time for someone to pick me. And I was lonely, so very very lonely. When a boy who already had a 3-book deal with a major publisher got the only writing grant available in the state to writers under 30, something finally snapped for me. I was sick of waiting; it was time to choose myself. I couldn't be rejected if I was the one creating the thing, right?

It was when I took the conscious decision to step off the beaten path that things changed for me. I created my own opportunities, but in a way that no one else was doing at the time – I created them so that I was making and creating WITH someone else. The power of collaboration runs through everything I do now, from the very first writing and publishing project I created in my little city of Adelaide, which spiralled into a 5-year international endeavour that would turn into the award-winning storytelling app, Story City, and lift up over 300 storytellers across half a dozen creative industries.

In creating my own opportunities, in making things like Story City, my novels, my branding work, I realised I made a place where I belonged, and where hundreds and thousands of others realised they belonged.

The success that I have had today is due largely to the power of story. Of how stories allow you to be understood for you, and to connect beyond yourself. I've won awards, presented hundreds of hours of storytelling workshops internationally, published 6 books, edited and/or published dozens of authors, I am a global entrepreneur of an app that helps you explore and connect to a city and the stories of its people, and I'm part of a 6 person team that brands a handful of high-flying femmpreneurs every year.

While much of that has been because of hard work, talent, and practice, the truth of the matter is I have gotten this far because I have chosen to make things together, rather than alone. To hone my understanding, skills and stories, with outside eyes, because through collaboration I make far more impact than I ever would on my own.

So I say to you pick yourself, don't wait for others to pick you. But also pick doing it together, rather than doing it alone.

Find your people. Band together. And you will make great things.

Contact Emily Online

Facebook: *http://www.facebook.com/EmilyCravenAuthor*

Website (bookmark me!): *http://www.cravenstories.com*

Twitter: *@cravenstories*

Instagram: *@imagesforjoy*

Work With Me: http://www.cravenstories.com/work-with-me/

E-Book Revolution Cover Design
By Neil Williams (Front Cover) & Kit Foster (Full Print Cover)

Email: kit@literartydesign.com

www.ingramcontent.com/pod-product-compliance
Lightning Source LLC
Chambersburg PA
CBHW032337300426
44109CB00041B/1089